GMDSS

for

NAVIGATORS

GMDSS
for
NAVIGATORS

P.C. Smith
P.M.G. 2nd Class Cert. (U.K.),   C.G.L.I. Telecomms. Certs. (U.K.),
R.O.G.C.P. Cert. (Aust.)   G.M.D.S.S. G.O.C. Cert. (Aust.)

Eur Ing J.J. Seaton
C. Eng.,   M.I.Mar.E.,   M.I.Mech.E.,   C.P.Eng.,   M.I.E. (Aust.)

Butterworth-Heinemann Ltd
Linacre House, Jordon Hill, Oxford OX2 8DP

℞ A member of the Reed Elsevier plc group

OXFORD LONDON BOSTON
MUNICH NEW DELHI SINGAPORE SYDNEY
TOKYO TORONTO WELLINGTON

First published 1994

British Library Cataloguing in Publication Data
A catalogue record for this book is available from the British Library

ISBN 0 7506 2177

Library of Congress Cataloguing in Publication Data
A catalogue record for this book is available from the Library of Congress

Composition by The Pinpoint Group, London
Printed and bound in Great Britain by Hartnolls

CONTENTS

Trickle Charge
Connecting Batteries
Ampere-Hours (AH)
Maintenance of Batteries
Specific Gravity
Safety
Uninterrupted Power Supplies (UPS)
Terminology
Fuses

PREFACE

The oil price booms of the nineteen-seventies served as the catalysts to change within many progressive industrial scenarios. The maritime industry is one such scenario.

Over the last two decades significant technological development has resulted from increased knowledge, understanding, and application of the discipline of engineering electronics. As ships and their equipment have paralleled such developments, it has led to major revisions in on-board working practices. Communication practices between vessels and the shore are changing markedly as electronic communications technology and its associated reliability progressively replaces the dedicated communicator on board ship.

The Global Maritime Distress and Safety System or GMDSS as it is abbreviated and referred to, is becoming the all-embracing term for ALL aspects of communication and data transfer between ship and shore/ship and ship. As the equipment and facilities to support GMDSS operation have developed, it has been necessary to also formulate the rules and regulations to accommodate the good working practices of both the operated and the operator

Consistent with maritime tradition, the governments of the seafaring nations of the world have assisted in establishing the rules with which they will all comply as GMDSS equipment is installed on new vessels and gradually replaces existing (and now) out-dated equipment of a bygone technological era.

This book brings together in one publication the knowledge required to assist anyone who seeks to fulfil the statutory requirements to become a qualified and competent operator of GMDSS communications equipment. The presentation of information within the book has purposely followed the Harmonised Examination Procedures adopted by the European Radiocommunication Committee (ERC)/Recommendation T/R 31-03E (Bonn 1993). (This Examination Syllabus may be found at Appendix 1). The Recommendation is based on the harmonised syllabi developed by the European Conference of Postal and Telecommunications Administrations (CEPT). Operator's Certificates will be recognized by all CEPT countries which have adopted Recommendation T/R 31-03E (Bonn 1993). The authors acknowledge the work of all parties responsible in finalising the requirements specified therein.

In conclusion it is the opinion of the authors that educators and others, who seek to train operators for this examination, will find the order of presentation of information within this publication complementary to the generation of courses designed for **competency-based curriculae**.

John Seaton/Pete Smith

GLOSSARY

AAIC	Accounting Authority Identification Code	**MERSSAR**	Merchant Ship Search And Rescue
AF	Audio Frequency	**MF**	Medium Frequency
AH	Ampere Hours	**MID**	Maritime Identification Digits
AM	Amplitude Modulation	**MMSI**	Maritime Mobile Service Identity
AMSA	Australian Maritime Safety Authority	**MRCC**	Maritime Rescue Co-ordination Centre
ARQ	Automatic Request Query		
ATU	Antenna Tuning Unit	**MSI**	Maritime Safety Information
AUSREP	Australian Ship Reporting System	**NAVTEX**	Single Frequency Time-Shared Broadcast System with Automated Reception and Message Rejection/Selection Facilities
CC	Coast Charge		
CES	Coast Earth Station		
CNID	Closed Network Identifier		
COSPAS	Space System for Search of Distress Vessels	**NBDP**	Narrow Band Direct Printing
		NCC	Network Control Centre
DCE	Data Circuit (Terminal) Equipment	**NCS**	Network Co-ordination Station
		NICAD	Nickel Cadmium
DNID	Data Network Identifier	**OSC**	On-Scene Commander
DSC	Digital Selective Calling	**R**	Resistance
DTE	Data Terminal Equipment	**RCC**	Rescue Co-ordination Centre
EGC	Enhanced Group Calling	**RF**	Radio Frequency
EPIRB	Emergency Position Indicating Radio Beacon	**RX**	Receive/Receiver
		SAR	Search And Rescue
FEC	Forward Error Correction	**SARSAT**	Search And Rescue Satellite Aided Tracking
FM	Frequency Modulation		
GF	Gold Franc	**SART**	Search And Rescue Radar Transponder/Survival Craft Radar Transponder
GMDSS	Global Maritime Distress and Safety System		
HF	High Frequency	**SC**	Ship Charge
H3E	Single Side Band Full Carrier	**SDR**	Special Drawing Right
I	Current	**SES**	Ship Earth Station
ID	Identification	**SHF**	Super High Frequency
IMO	International Maritime Organization	**SOLAS**	Safety Of Life At Sea
		SSB	Single Side Band
INMARSAT	International Maritime Satellite Organization	**SSFC**	Sequential Single Frequency Code
IRS	Information Receiving Station	**TOR**	Telex Over Radio
ISS	Information Sending Station	**TX**	Transmit/Transmitter
ITU	International Telecommunications Union	**UHF**	Ultra High Frequency
		UPS	Uninterrupted Power Supply
J3E	Single Side Band Suppressed Carrier	**USB**	Upper Side Band
		V	Volts
LF	Low Frequency	**VDU**	Visual Display Unit
LL	Land Line	**VHF**	Very High Frequency
LSB	Lower Side Band	**VLF**	Very Low Frequency
LUT	Local User Terminal	**W**	Watts

Part 1

Knowledge of the Basic Features of the Maritime Mobile Service and the Maritime Mobile Satellite Service

THE GENERAL PRINCIPLES AND BASIC FEATURES OF THE MARITIME MOBILE SERVICE

TYPES OF COMMUNICATION IN THE MARITIME MOBILE SERVICE

Distress, Urgency, and Safety Communications traffic consists of all messages relating to the immediate assistance required by the ship in DISTRESS, including Search And Rescue (SAR) communications and on-scene communications. Distress traffic must be followed by all stations, even if they are not assisting, until it is clear that assistance is being rendered.

Public Correspondence is correspondence between ship stations and shore stations of a commercial nature, for example radiotelegrams and radiotelephone calls.

Port Operations Service is a maritime mobile service, in or near a port, between coast stations and ship stations, or between ship stations, in which messages are restricted to those relating to the operational handling, the movement and the safety of ships and, in emergency, to the safety of persons.

Ship Movement Service is a safety service in the maritime mobile service other than a port operations service, between coast stations and ship stations, or between ship stations, in which messages are restricted to those relating to the movement of ships.

Intership Communications embracing navigation and safety communications are those VHF radiotelephone communications conducted between ships for the purpose of contributing to the safe movement of ships.

On-board Communications are low powered communications in the maritime mobile service intended for use for internal communications on board a ship, or between a ship and its lifeboats and life rafts during lifeboat drills or operations, or for communication within a group of vessels being towed or pushed, as well as for line handling and mooring instructions.

TYPES OF STATION IN THE MARITIME MOBILE SERVICE

Ship Station
A mobile station in the maritime mobile service located on board a vessel which is not permanently moored, other than a survival craft station.

Coast Station
A land station in the maritime mobile service.

Pilot Stations, Port Stations
Coast stations in the port operations service.

Aircraft Station
A mobile station in the aeronautical mobile service, other than a survival craft station, located on board an aircraft.

Rescue Coordination Centre (RCC), as defined in the International Convention on Maritime Search and Rescue 1979, refers to a unit responsible for promoting the efficient organisation of search and rescue services and for coordinating the conduct of search and rescue operations within a search and rescue region. These units may also be known as Maritime Rescue Coordination Centres (MRCCs).

The First RCC is the RCC affiliated with the coast station which received the Distress Alert, and which should then take responsibility for co-ordinating SAR operations. However, if there is confusion over which station is the First RCC because an Alert has been acknowledged by more than one station, then the RCCs must agree between themselves which is the First RCC.

ELEMENTARY KNOWLEDGE OF FREQUENCIES AND FREQUENCY BANDS

The Concept of Frequency
A transmitter produces a wave composed of electrical energy and magnetic force. These electromagnetic waveforms leave a transmitter antenna at the speed of light, and the number of waveforms (called Hertz) which leave the antenna in 1 second, give us the frequency. The greater the number of waveforms leaving the antenna, the higher the frequency. Radio frequencies emanate naturally from certain materials

and we can even say they have become a natural resource; for example, radio, television, and microwave ovens all use radio waves at different frequencies. These waveforms are known as Hertzian waves, or Hertz (Hz) and may be grouped together so that a table of frequencies can be shown. This table of frequencies is called the **Radio Frequency Spectrum**.

The Equivalence Between Frequency and Wavelength

Each waveform has a physical length called wavelength, which is inversely proportional to the frequency. So the higher the frequency, the shorter the wavelength. Wavelength can be easily calculated by the formula:-

Wavelength = **Frequency** divided by the **speed of light**. So a frequency of **2,000,000 Hz** divided by **300,000,000** metres per second would give us a **wavelength** of **150 metres**.

The Radio Frequency Spectrum

FREQUENCY BANDS		FREQUENCIES
Super High Frequencies	SHF	3 GHz-30 GHz
Ultra High Frequencies	UHF	300 MHz -3 GHz
Very High Frequencies	VHF	30 MHz -300 MHz
High Frequency	HF	3 MHz-30 MHz
Medium Frequency	MF	300 kHz-3 MHz
Low Frequency	LF	30 kHz-300 kHz
Very Low Frequency	VLF	3kHz-30kHz

The Unit of Frequency is the Hertz. One Hertz is one cycle of energy or one Hertzian wave and as Metric notation is used, so one kiloHertz (**kHz**) is equal to one thousand Hertz, one MegaHertz (**MHz**) equals one million Hertz, and one GigaHertz (**GHz**) equals one thousand million Hertz.

The Subdivision of the Most Significant Parts of the Radio Frequency Spectrum: MF – HF – VHF – UHF – SHF

Frequencies are separated into Bands to form what is known as the Radio Frequency Spectrum. This is usually a table showing all radio frequencies progressively from the lowest to the highest. Frequencies within each band are allocated by the **International Telecommunications Union (ITU)** for different uses.

MF band is **300 kHz- 3 MHz**   **HF** band is **3 MHz – 30 MHz**
VHF band **is 30 MHz – 300 MHz**   **UHF** band is **300 MHz – 3 GHz**
                        SHF band is **3 GHz – 30 GHz**

CHARACTERISTICS OF FREQUENCIES

Different Propagation Mechanisms: Propagation in Free Space, Ground Wave, Ionospheric Propagation

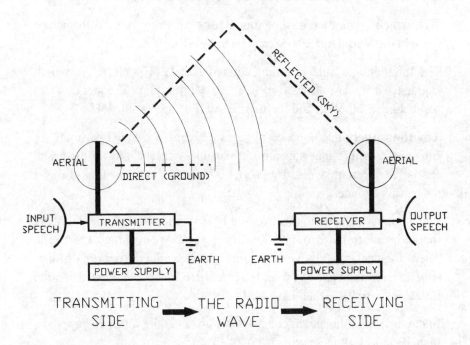

Communications Systems

All communications systems must have the same basic component parts. There must be a power supply, a transmitter, an aerial (or antenna) and a receiver. All equipment must be earthed. A transmitter and a receiver in the same casing is called a transceiver.

When we speak into a microphone, our voice is converted from sound waves or **A**udio **F**requencies (**300 Hz – 3 kHz or AF**), to electrical energy, that is, voltage and current.

The transmitter produces a wave of electromagnetic energy which we call a carrier wave, and which then carries our voice energy to the antenna, to be radiated.

Modulation

The process of 'superimposing' a voice onto a carrier wave is called MODULATION, and there are two methods in marine radio of modulating carrier waves for Radiotelephony. These are called AMPLITUDE MODULATION – known as **AM**, and FREQUENCY MODULATION, known as **FM**.

AM is used on all marine transmissions from 2182 kHz up to 30 MHz, and is known generally as **SSB** (or **S**ingle **S**ide **B**and).

FM, which is another way of modulating a carrier, is used in marine radio in the **V**ery **H**igh **F**requency band (**VHF**).

The **I**nternational **T**elecommunications **U**nion (**ITU**) determines which frequencies we are allowed to use, the Marine High Frequency (HF) band being 3-30 MHz and the marine VHF band 30-300 MHz.

The Ionosphere is composed of several layers of charged particles -or Ions – (ie atomic and sub atomic particles which hold an electrical charge) at different levels, from about 60 km to more than 500 km above the earth's surface.

The amount of charge in the different layers varies all the time; there are marked variations through morning, afternoon, evening, and night. There are weekly, monthly, and annual changes, and the pattern of Ionospheric radiation has been found to exactly follow the incidence of solar flares or sunspots, which have an 11 year cycle.

In daytime the Ionosphere consists of 4 layers at the following approximate distances above earth:-

the **D** layer about 60-80 km,

the **E** Layer about 80-160 km,

the **F1** layer about 160-210 km,

the **F2** layer about 210 km to more than 500 km.

The intensity of radiation – which is always varying – has profound effects on radio wave propagation, and the most marked changes we must take account of, are from day to night.

Generally, we can say that high frequencies are best during daylight hours, and low frequencies best during hours of darkness.

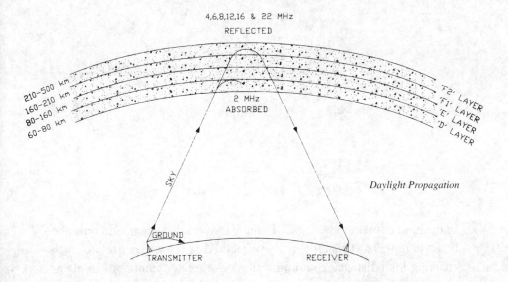

Daylight Propagation

Propagation of MF Frequencies
Each transmitted wave has two components, or parts; the ground wave and the sky wave.

Ground waves tend to follow the curvature of the earth and decay (lose energy) very rapidly. Consequently ground waves have a comparatively short range of around 100 km during daylight hours. The band of frequencies from 300-3000 kHz is the MF band which tends to use ground waves only, because sky waves in this band are absorbed by the D layer.

The AM broadcast band 530-1602 kHz has a short range for the above-mentioned reason i.e. its sky waves are all absorbed by the D layer during daylight hours so that only the ground waves are propagated.

Propagation of Different HF Frequency Bands
Long distance communication is achieved by using sky waves, which radiate up towards space and are then refracted (turned back to earth) by

the Ionosphere, to be received at a distant receiver antenna. Sky waves may complete several 'skips' covering many thousands of miles thus enabling communications over great distances.

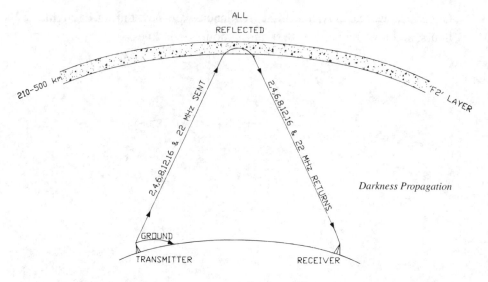

Darkness Propagation

In the hours of darkness, the D, E and F1 layers disappear, and only the F2 layer remains. This effectively refracts ALL sky waves back to earth enabling long distance communication on a greater number of frequencies, including lower frequencies such as 2182 kHz. There is also greatly increased interference on all frequencies.

Remember, that because the Ionosphere is always changing, different frequencies will be refracted (or reflected) back to earth at different times of day or night, so efficient communication by radio is largely a matter of experience.

Propagation of VHF and UHF Frequencies

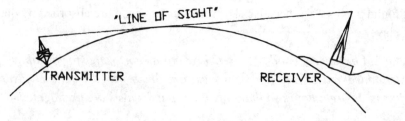

'Line of Sight' transmission

On VHF and UHF the Ionosphere is not used because wavelengths at these frequencies are too short to be turned back towards earth. So we effectively have only a ground wave at VHF and UHF, which we call **'Line of Sight'** transmission.

Line of Sight means essentially that if the radio path between transmitter and receiver is unobstructed (by for example, a high land mass or building), then good communication should be achieved.

VHF offers short range high quality communication of up to say 30 km ship to ship, and about 60 km ship to shore. The greater ship-shore range is because shore stations have much higher antennae than ships, therefore, the higher the antenna the greater the range.

KNOWLEDGE OF THE ROLE OF THE VARIOUS MODES OF COMMUNICATION

Digital Selective Calling (DSC) is a method of automatically alerting ship or shore stations, individually or collectively.

Individually, by dialling the **M**aritime **M**obile **S**ervice **I**dentity (**MMSI**) of the required station into the DSC controller.

Collectively, by selection of the **All Ship** format, or when in Distress, by pressing the **Distress** pad.

A good example of DSC usage would be a **Safety** call addressed to **All Ships** and specifying a radiotelephone frequency for subsequent broadcast of, say, a navigation warning. Ships receiving the call would then adjust their MF/HF or VHF transceivers for reception of the warning.

The DSC controller is permanently connected to the ship's main transceiver, and enables digitally coded signals to be transmitted on specific frequencies. The controller should also be interfaced with the ship's navigation system so that the ship's position which is continually updated, will always be correct in the event of, for example, a Distress Alert transmission.

Digital calls always include the identity (**MMSI**) of the transmitting ship and a signal which when received, will activate aural and visual alarms. In this respect, the DSC controller could be thought of as a 'pager' which is used to alert ships and coast stations prior to subsequent voice or radiotelex communications.

When not being used for transmission, the controller will always be in **receive** mode in order to **SCAN** (watch) certain DSC frequencies through the ship's transceiver ie. a watch on VHF Channel 70, or on MF 2187.5 kHz. If the vessel trades in A3 or A4 areas and does not have a Ship Earth Station, then a watch on 2187.5 kHz, 8414.5 kHz and one other frequency in the band in which the ship is working must be kept.

Watch should also be kept on frequencies used by a DSC coast station for notification of eg automatic or semi-automatic telephone calls. It is also possible for ships to initiate automatic or semi-automatic telephone calls by selectively calling an individual coast station and including a working channel and telephone number in the call. Coast station equipment then processes the call through national or international telephone networks so that both subscribers are connected with minimal or no intervention by coast operators.

The DSC call **always** includes the transmitting station identification (MMSI), and for **Individual** or **Single Ship** calls, the MMSI of the called station.

Radiotelephony
Radio: A general term applied to the use of radio waves.
Telephony: A form of telecommunication set up for the transmission of speech or, in some cases, other sounds.
Radiotelephone Call: A telephone call, originating in or intended for a mobile station or a mobile earth station, transmitted on all or part of its route over the radio communication channels of the mobile service or of the mobile-satellite service.

Narrow Band Direct Printing (NBDP) Telegraphy
A service also known as radiotelex, and used in communications via satellite, HF radio, and MF radio, for example, International NAVTEX.

Facsimile
A form of telegraphy for the transmission of fixed images, with or without half-tones, with a view to their reproduction in a permanent form.

Data
A Data Call is one requested for the purpose of exchanging data of any kind between telephone stations specially equipped to transmit and receive such data. Data (the reduction of information to 'bits' ie 1 or 0) may be transmitted and received using the INMARSAT satellite system

by stations with the appropriate equipment. Data transmissions also emanate from buoy stations used for oceanographic data transmission and by stations interrogating these buoys.

Morse Telegraphy

Telegraphy: A form of telecommunication which is concerned in any process providing transmission and reproduction at a distance of documentary matter, such as written or printed matter or fixed images, or the reproduction at a distance of any kind of information in such a form. For the purposes of the Radio Regulations, unless otherwise specified therein, telegraphy shall mean a form of telecommunication for the transmission of written matter by the use of a signal code – e.g. the Morse Code.

Radiommunication by the use of Morse Code is still practised by many ships using the HF band of frequencies.

Morse is not a requirement of the GMDSS.

ELEMENTARY KNOWLEDGE OF DIFFERENT TYPES OF MODULATION AND CLASSES OF EMISSION

Classes of Emission

Emission: Radiation produced, or the production of radiation, by a radio transmitting station.
Classes of emission, are the sets of characteristics of an emission, designated by standard symbols, eg. type of modulation of the main carrier, modulating signal, type of information to be transmitted and also, if appropriate, any additional signal characteristics.

Carrier Frequency and Assigned Frequency

Carrier Frequency or Characteristic Frequency is a frequency which can be easily identified and measured in a given emission.

Assigned Frequency is the centre of the frequency band assigned to a station.

Bandwidth of Different Emissions

Necessary Bandwidth for a given class of emission, is the width of the frequency band which is just sufficient to ensure the transmission of information at the rate and with the quality required under specified conditions.

Official Designations of Emissions

Emissions shall be designated according to their necessary bandwidth and their classification e.g.

F1B **F** means: Frequency Modulation or **FM**.
  1 means: A single channel containing quantized or digital information without the use of a modulating sub-carrier.
  B means: Telegraphy – for automatic reception.
F1B in Marine Radio is used for **HF** Radiotelex and **DSC**.

F3E **F** means: Frequency Modulation or **FM**.
  3 means a single channel containing analogue information.
  E means Telephony (including sound broadcasting).
F3E is used in Marine Radio on **VHF** Radiotelephony eg. Channel 16.

J3E **J** means: Single-Side-Band, **(SSB)** with a suppressed carrier.
  3 means: A single channel containing analogue information.
  E means: Telephony (including sound broadcasting).
J3E is used on **ALL** Radiotelephony frequencies in the Marine **HF** band (3-30 MHz).

A3E **A** means: Double-Side-Band **(DSB)**.
  3 means: A single channel containing analogue information.
  E means: Telephony (including sound broadcasting).
A3E is used by all **MF** radio broadcast stations.

A1A **A** means: Double-Side-Band **(DSB)**.
  1 means: A single channel containing quantized or digital information without the use of a modulating sub-carrier.
  A means: Telegraphy – for aural reception.
A1A is used for **HF** Morse telegraphy.

H3E **H** means: Single-Side-Band **(SSB)** with full carrier.
  3 means: A single channel containing analogue information.
  E means: Telephony (including sound broadcasting).
H3E is used on 2182 kHz for Distress and Radiotelephone Alarm signal broadcasts.

R3E **R** means: Single-Side-Band **(SSB)** with a reduced or variable level carrier
  3 means: A single channel containing analogue information
  E means: Telephony (including sound broadcasting)
R3E is not used in Marine radiocommunications.

J2B   **J** means: Single-Side-Band (**SSB**) with a suppressed carrier
2 means: A single channel containing quantized or digital
information with the use of a modulating sub-carrier
B means: Telegraphy – for automatic reception
J2B is used for Radiotelex in Marine radiocommunications.

Unofficial Designations of Emissions

F1B and J2B   **known as: TELEX or RADIOTELEX**
J3E   known as: SSB
A3E and H3E known as: AM
A1A known as: CW

FREQUENCIES ALLOCATED IN THE MARITIME MOBILE SERVICE

The Usage of MF, HF, VHF, UHF, and SHF Frequencies in the Maritime Mobile Service

SOME EXAMPLES OF FREQUENCY USAGE		
FREQUENCY	**BAND**	**USAGE**
9 GHz	SHF	X-Band Radar
9 GHz		SART
3 GHz		S-Band Radar
1.6 GHz	UHF	INMARSAT
1.6 GHz		'L' Band EPIRBs
406 MHz		COSPAS-SARSAT
156.8 MHz	VHF	Channel 16
156.525 MHz		DSC Channel 70
121.5 MHz		EPIRBs
8291 kHz	HF	R/Telephony Distress
8414.5 kHz		DSC Distress
5680 kHz		Ship-Aircraft SAR
2182 kHz	MF	R/Telephony Distress
2187.5 kHz		DSC Distress
518 kHz		NAVTEX
285 kHz	LF	RDF Beacons
130 kHz		DECCA
100 kHz		LORAN-C
14 kHz		VLF OMEGA

The Concept of Radio Channel. Simplex, Semi-Duplex, and Duplex. Paired and Unpaired Frequencies.

Radio channels may be thought of as frequencies which have been designated for specific purposes eg Radiotelephony, Radiotelex, etc. The channels may consist of paired frequencies and will be known by a Channel Number allocated by the International Telecommunications Union (ITU).

The first digit or the first two digits of the channel number, indicate the frequency band, the remaining digits are the actual channel number. For example in Channel **1602: 16** means the frequencies are located in the 16 MHz band and **02** means it is the second channel within that band.

Channel 1602 contains the ship receive frequency 17245 kHz and ship transmit frequency 16363 kHz.

In Channel **404: 4** indicates the frequencies are in the 4 MHz band and **04** indicates it is the second channel within that band.

Channel 404 contains the ship receive frequency of 4366 kHz and the ship transmit frequency of 4074 kHz.

Unpaired frequencies may also be selected for communication (including cross-band) from the relevant Appendices noted on page 15.

Simplex Operation refers to the method in which transmission is made possible alternately in each direction of a telecommunication channel, for example, by means of manual control. Simplex is used on all marine Distress and Calling frequencies such as 2182 kHz, when both stations use the same frequency for transmit and receive. Therefore only one station can transmit at a time and indicates the end of a transmission by use of the word 'Over'.

Semi-Duplex Operation is a method which is simplex operation at one end of the circuit and duplex operation at the other.

Duplex Operation is the operating method in which transmission is possible simultaneously in both directions of a telecommunication channel. Duplex is used on all marine radiotelephone channels, both stations transmitting on separate frequencies, so that a normal conversation may take place.

Frequency Plans and Channelling Systems

Frequency and channel allocations may be found in the following tables:-

HF Telephony	See Appendix 16 of the Radio Regulations.
VHF Telephony	See Appendix 18 of the Radio Regulations.
HF NBDP	See Appendices 32 & 33 of the Radio Regulations.

MF Telephony and NBDP – Additional Provisions applying to Region 1

All radiotelephone stations on ships making international voyages should be able to use the following ship-shore working frequencies if required by their service: 2045 kHz for J3E emissions and the intership frequency 2048 kHz for J3E emissions if also required by their service. This latter frequency may be used as an additional ship-shore frequency.

These frequencies shall not be used for working by stations of the same nationality. Ships frequently exchanging correspondence with a coast station of a nationality other than their own may use the same frequencies as ships of the nationality of the coast station where:

a) mutually agreed by the administration concerned; and

b) where the facility is open to ships of all nationalities by virtue of a note against each of the frequencies concerned in the List of Coast Stations.

The following ship-shore frequencies may be assigned to coast stations as receiving frequencies:

2051 kHz, 2054 kHz and 2057 kHz.

Frequency for International NAVTEX (MF NBDP) is 518 kHz.

For additional information on Frequency Planning and Channelling arrangements for countries in Region 1 see the *Final Acts of the Regional Administrative Radio Conference for the Planning of the MF Maritime Mobile and Aeronautical Radionavigation Services (Region 1) Geneva 1985*).

See Manual for Use by the Maritime Mobile and Maritime Mobile-Satellite Services for Regions map.

GMDSS Distress and Safety Frequencies

RADIO-TELEPHONY	DSC	NBDP	NBDP/MSI
			490 kHz
			518 kHz NAVTEX
2182 kHz	2187.5 kHz	2174.5 kHz	
3023 kHz			
4125 kHz	4207.5 kHz	4177.5 kHz	4209.5kHz NAVTEX
			4210 kHz
5680 kHz			
6215 kHz	6312 kHz	6268 kHz	6314 kHz
8291 kHz	8414.5 kHz	8376.5 kHz	8416.5 kHz
12290 kHz	12577 kHz	12520 kHz	12579 kHz
16420 kHz	16804.5 kHz	16695 kHz	16806.5 kHz
			19680.5 kHz
			22376 kHz
			26100.5 kHz

156.3 MHz   (VHF Channel 6 Intership)

156.525 MHz   (VHF Channel 70 DSC Alerting)

156.650 MHz   (VHF Channel 13 Intership MSI)

156.8 MHz   (VHF Channel 16 Distress, Safety & Calling)

Allocated Frequency Bands
406-406.1 MHz
Used exclusively by satellite EPIRBs in the Earth-Space direction.

1530-1544 MHz
Available for routine non-safety purposes and also used for Distress and Safety purposes in the space-Earth direction in the maritime mobile satellite service.

1544-1545 MHz
Use of this band (Space-Earth) limited to Distress and Safety operations including:

a) feeder links of satellites needed to relay the emissions of satellite EPIRBs to earth stations; and

b) narrow-band (Space-Earth) links from space stations to mobile stations.

1626.5-1645.5 MHz
Available for routine non-safety purposes and is used for Distress and Safety purposes in the Earth-space direction in the maritime mobile satellite service.

1645.5-1646.5 MHz
Use of this band (Earth-space) is limited to Distress and Safety operations including:

a) transmissions from satellite EPIRBs; and

b) relay of distress alerts received by satellites in low polar earth orbits to geostationary satellite.

9200-9500 MHz
Used by radar transponders to facilitate search and rescue.

Survival Craft Stations
1) Equipment for radiotelephony use in survival craft stations shall, if capable of operating on any frequency in the bands between 156 MHz and 174 MHz, be able to transmit and receive on 156.8 MHz and at least one other frequency in these bands.

2) Equipment for transmitting locating signals (SARTs) from survival craft stations shall be capable of operating in the 9200-9500 MHz band.

3) Equipment with digital selective calling facilities for use in survival craft shall, if capable of operating in the bands between:
   a) 1605 kHz and 2850 kHz, be able to transmit on 2187.5 kHz;
   b) 4000 kHz and 27500 kHz, be able to transmit on 8414.5 kHz; and
   c) 156 MHz and 174 MHz, be able to transmit on 156.525 MHz.

Distress and Safety Frequencies of the Pre-GMDSS System

MORSE	RADIOTELEPHONY
500 kHz	
	2182 kHz
	3023 kHz
	4125 kHz
	5680 kHz
	6215.5 kHz
8364 kHz	
	121.5 MHz & 123.1 MHz
	156.3 MHz & 156.8 MHz

Survival Craft Stations

Equipment provided for use in survival craft stations shall, if capable of operating on any frequency in the bands between:

1) 405 kHz and 535 kHz, be able to transmit with a carrier frequency of 500 kHz using either class A2A and A2B or H2A and H2B emissions. If a receiver is provided for any of these bands, it shall be able to receive class A2A and H2A emissions on a carrier frequency of 500 kHz;

2) 1605 kHz and 2850 kHz, be able to transmit with a carrier frequency of 2182 kHz using class A3E or H3E emissions. If a receiver is provided for any of these bands, it shall be able to receive class A3E and H3E emissions on a carrier frequency of 2182 kHz;

3) 4000 kHz and 27500 kHz, be able to transmit with a carrier frequency of 8364 kHz using class A2A or H2A emissions. If a receiver is provided for any of these bands, it shall be able to receive class A1A, A2A and H2A emissions throughout the band 8341.75 – 8728.5 kHz;

4) 118 MHz and 136 MHz, be able to transmit on 121.5 MHz, preferably using amplitude modulated emissions. If a receiver is provided for any of these bands, it shall be able to receive class A3E emissions on 121.5 MHz;

5) 156 MHz and 174 MHz, be able to transmit on 156.8 MHz using class G3E emissions. If a receiver is provided for any of these bands it shall be able to receive class G3E emissions on 156.8 MHz; and

6) 235 MHz and 328.6 MHz, be able to transmit on the frequency 243 MHz.

Calling Frequencies

Ship Stations May Use The Following Carrier Frequencies For Calling In Radiotelephony	Coast Stations May Use The Following Carrier Frequencies For Calling In Radiotelephony
	4125 kHz
	4417 kHz
4125 kHz	6215 kHz
6215 kHz	6516 kHz
8291 kHz	8291 kHz
8255 kHz	8779 kHz
12290 kHz	13137 kHz
16420 kHz	17302 kHz
18795 kHz	19770 kHz
22060 kHz	22756 kHz
25097 kHz	26172 kHz

Notes
1) 8291 kHz reserved exclusively for Distress.
2) 4125 kHz, 6215 kHz, 12290 kHz and 16420 kHz also used for Distress and Safety traffic.

The General Principles and Basic Features of the Maritime Mobile Satellite Service

BASIC KNOWLEDGE OF SATELLITE COMMUNICATIONS

INMARSAT Space Segment

This consists of the satellites and the radio communications transponders and telemetry and tracking devices which they carry. Four active satellites some 36,000 km above the equator between them cover the entire Earth's surface except for the Polar regions. The satellites are in geo-synchronous orbits and always 'see' the same area. Thus there is one satellite for the Pacific Ocean Area, one satellite for the Indian Ocean Area and two satellites for Atlantic Ocean Area coverage. The Space Segment also includes ground control facilities.

Modes of Communication

Telex Services are available in ship-shore and shore-ship directions and are provided on INMARSAT-A, INMARSAT-B and INMARSAT-C stations.

Telephone Services are available in ship-shore and shore-ship directions and are provided on INMARSAT-A and INMARSAT-B stations.

Data and Facsimile Communications

Data communications are available in ship-shore and shore-ship directions using INMARSAT-A, INMARSAT-B and INMARSAT-C equipment.

Facsimile communications are available in the ship-shore direction on INMARSAT-A, INMARSAT-B and INMARSAT-C equipment.

However, facsimile communications in the shore-ship direction are not available on INMARSAT-C and are available on INMARSAT-A and INMARSAT-B only with an extra modem connected.

Store and Forward Operation

Based on digital technology, this is the method used in the transmission of telex and data information. A ship to shore message is transmitted as 'packets' of data and stored at the Coast Earth Station (CES). When the CES receives errors, it repeatedly requests re-transmission until the information has been correctly received. The message format is then converted into one which can be forwarded by the requisite telecommunications network. e.g. telex. In the shore to ship direction, messages are stored at the CES then converted into the correct format for delivery to the ship.

Distress and Safety Communications are available on all INMARSAT terminals. The INMARSAT (International Maritime Satellite Organisation) system provides immediate and routine access to Search And Rescue (SAR) services.

INMARSAT-A Communications Services cover the full range of user requirements ie Telephony (direct dial), Telegram, Telex and Telex Group Call Service, Medium and High Speed Data Service and Access to Data Banks, Facsimile, Subscription News Service, Voice/Data Group Call Service, Compressed Video Television Service, International Packet Switching Service, Voice Messaging Service, Radiopaging Service, Videotex Service, Electronic Mail Services, Telex Delivery by Telephone Service, Telex Delivery by Mail Service, Credit Card Telephones, Credit Access.

All of the above Services are available on **INMARSAT-B**.

INMARSAT-C Communications Services
Digital technology providing Telex, Telegram, and Data Services, including Switched Packet Data Network Service.

INMARSAT Enhanced Group Call (EGC) System
EGC is a system of automatic access in the shore-ship direction only which enables delivery of printed messages to individual ships, or groups of ships (whether grouped by fleet, flag, or geographical area). The system ensures automatic reception of Maritime Safety Information, or group messages to Fleets.

TYPES OF STATION IN THE MARITIME MOBILE SATELLITE SERVICE.

Coast Earth Station (CES)
Allows ship-shore and shore-ship communication via satellite. Mobile stations can select the CES they wish to use.

Network Coordination Station (NCS)

One NCS in each Ocean Area controls operations of all CESs within its area, and is always linked to its CESs for control purposes.

The NCS transmits on a Common TDM Channel (Time Division Multiplex Channel or Frequency) which is received by all Ship Earth Stations when they are not busy. *The Common Channel is used by NCS to broadcast EGC messages, and also for certain data transfer operations.*

The NCS maintains a database (which includes location of mobile stations), synchronises their access to CESs and monitors all activities within its own region. All NCSs are linked for control purposes.

Note that all of the above functions are carried out automatically by SES, NCS and CES without any intervention by the SES operator.

Ship Earth Station (SES)

The commissioned INMARSAT terminal on board.

Network Control Centre (NCC) This station at INMARSAT headquarters in London, monitors, coordinates, and controls all NCSs to which it is permanently connected.

INMARSAT SES-NCS-CES Sequencing

1) SES tunes to the NCS Common Channel when idle

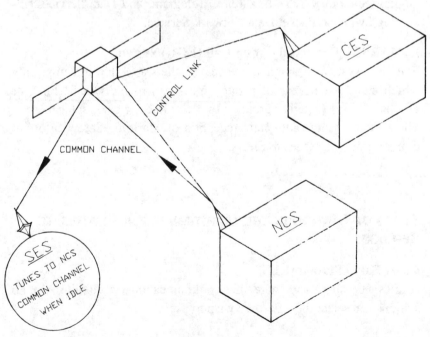

2) **SES** *calls* **NCS** *on a* **Signalling Channel** *which is used only for initial contact.* **SES** *requests call to* **CES**

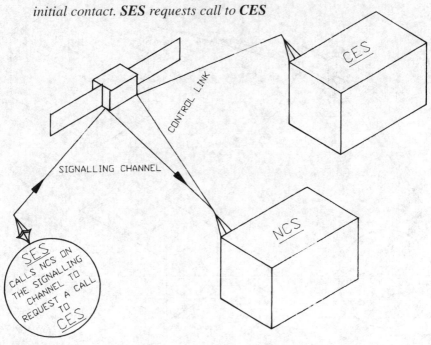

3) **NCS** *assigns* **SES** *to* **CES Message Channel**

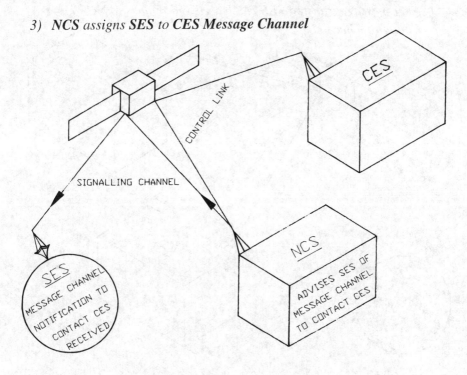

*4) **SES – CES** message transfer occurs*

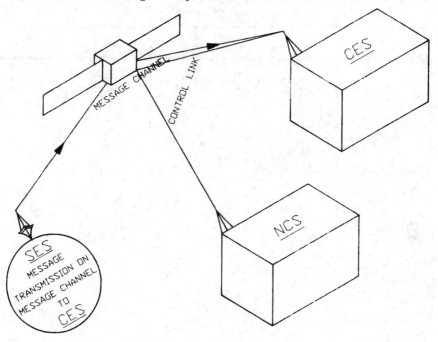

*5) After message transfer, **SES** is released and retunes to **NCS** **Common Channel***

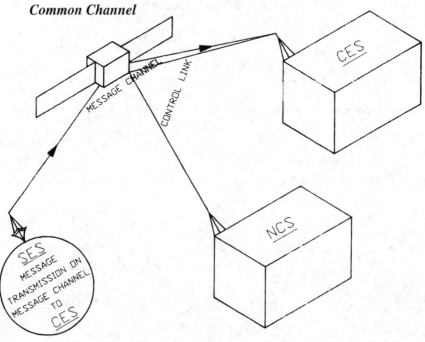

6) *Delivery of message confirmed to **SES** by **CES** via **NCS** on request (INMARSAT-C only)*

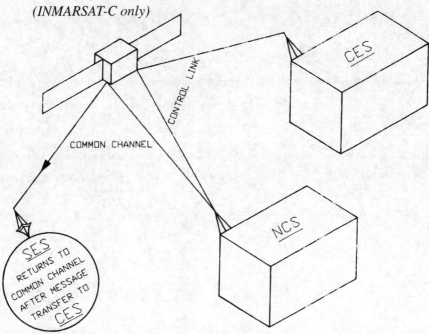

CONTROL LINK

COMMON CHANNEL

CES

NCS

SES
RETURNS TO
COMMON CHANNEL
AFTER MESSAGE
TRANSFER TO
CES

Part 2

Detailed Practical Knowledge and Ability To Use the Basic Equipment of a Ship Station

Knowledge of, and Ability to Use in Practice, the Basic Equipment of a Ship Station

WATCHKEEPING RECEIVERS

The Controls and Usage of the 2182 kHz Watch Receiver

ON/OFF   switch may be combined with the Volume (Gain) control.

Lamp   illuminated when equipment switched ON.

MUTE   prevents background noise, static, etc when ON.

TEST   simulates reception of Radiotelephone Alarm signal.

RESET   resets receiver to watchkeeping mode with MUTE on
RESET

USAGE   receiver is tuned to 2182 kHz, normally with MUTE on. The receiver is activated and its gain will automatically increase, when the Radiotelephone Alarm signal is received.

The Controls and Usage of VHF DSC Watch Receiver

ON/OFF   switch may be combined with a Gain control.

Lamp   illuminated when equipment switched ON.

TEST   runs internal test on Receiver.

USAGE   Receiver will activate aural and visual indications on receipt of a DSC Alarm on Channel 70.

The Controls and Usage of MF DSC Watch Receiver and MF/HF Watch Receiver

ON/OFF   switch

Lamp   illuminated when receiver switched ON.

Lamp   illuminated and flashing on receipt of DSC distress alert.

KEYPADS   used to make selections from a number of menus eg. frequencies to be scanned.

USAGE   receiver scans the MF and HF DSC distress frequencies as selected. On receipt of a DSC distress alert, will activate with aural and visual indications.

VHF Radio Installation

Typical DSC VHF equipment controls as below

Channels

The ITU has allocated 55 channel for Marine VHF use

Channel
70 – DSC alerting.
16 – International Distress and Safety.
13 – intership navigational information.
6 – primary intership channel.

Controls and Usage

POWER ON/OFF	may be combined with Gain control.
DISPLAY	illuminates when equipment switched on.
KEYPADS	used to select channels, or make selections from file memory menus.
SQUELCH	adjusted to minimum threshold so that background noise/static is silenced yet weak signals will activate receiver.
DW	Dual Watch pad. Enables receiver to watch channel 16 and one other channel. Will automatically select channel 16 in the event of a call on that channel.
CH16	selects channel 16.
25/1	power reduction 25 watts or 1 watt.

DSC

CH70	selects DSC panel.
ALARM + CH70	both these pads pressed at the same time to transmit DSC distress alert.
ALM.OFF	pressing this pad stops alarm call or other general call.
CALL	this pad causes selected memory file to be transmitted
* + CALL	two pads pressed at the same time causes a selected memory File to be transmitted 5 times.
ALM/DISABLE	pressing this pad disables audio alarms except distress/urgency calls.

MEMO 1	input time & position (for received calls addressed to area).
MEMO 2	input telephone numbers.
FILE	access file memories when pre-programming calls.
ENTER	enters information to FILE or MEMO.
FILES 1-15	will store transmitted calls.
FILES 16-20	will store received ordinary calls.
FILES 21-40	will store received distress calls.
#	clears mode and resets for DSC operation.

MF/HF Radio Installation

Typical marine radio transceiver controls as below

Frequencies:

Bands covered are 1.6 MHz to 30 MHz. Frequencies are fully synthe-sized and may be selected by pressing keypads.

Typical Controls and Usage:

Connecting the Power. This type of installation draws its supply from the ship's mains via a power supply unit, (or from the ship's batteries if using the emergency supply). Power switch on supply unit must be **ON**.

MAIN SWITCH   connects power to equipment.
SUPPLY ON/OFF switches equipment ON or OFF.

Selecting Receive (Rx) Frequency

PRESS PAD MARKED RX
SELECT NUMERAL KEYPADS EG 6215 kHz
PRESS PAD MARKED ENTER

Selecting Transmit (Tx) Frequency

PRESS KEYPAD MARKED TX
SELECT KEYPADS EG 6215
PRESS PAD MARKED ENTER

Selecting ITU Channel Number

SELECT PAD MARKED RCL
  (this recalls information from memory)
SELECT NUMERAL KEYPADS EG 1203
PRESS PAD MARKED ENTER

The ship transmit frequency 12236 kHz and the ship receive frequency 13083 kHz now appear in the display.

Tuning the Transmitter:
Most modern transmitters are fitted with an automatic tuning unit. If a manual tuning unit is used however, the following procedure should be adopted:

SELECT LOW POWER

SELECT TX FREQUENCY AND LISTEN FIRST TO CHECK
   IT IS CLEAR

SELECT AERIAL COIL TAP ACCORDING TO CHART FOR
   FREQUENCY

TURN TUNE AND LOAD CONTROLS TO ZERO

PRESS PAD MARKED TUNE

ADJUST LOAD THEN TUNE CONTROLS FOR HIGHEST
   READING POSSIBLE ON AERIAL TUNING UNIT.

Selecting the Class of Emission
Transmitters have several 'Modes' of emission eg J2B or F1B for Radiotelex, H3E (which is normally selected automatically) for a full carrier transmission on 2182 kHz, J3E for Radiotelephony on all frequencies. Other modes may be available on the transmitter, for example; R3E for a Single Sideband Reduced carrier, or LSB meaning Lower Sideband – only Upper Sideband or USB is used in Marine Radiotelephony.
Be sure to select the correct Mode for the desired emission.

Using Volume Control and Squelch

TURN SQUELCH UNTIL BACKGROUND NOISE/STATIC IS
   HEARD

TURN UP VOLUME

TURN UP SQUELCH UNTIL BACKGROUND NOISE
   DISAPPEARS

DO NOT TURN SQUELCH TOO FAR, OR ALL RECEIVED
   SIGNALS WILL BE LOST

Using Clarifier or Rx Fine Tuning

Received voice signals which are even slightly off frequency will be distorted. Slowly turning the CLARIFIER or FINE TUNE will compensate for frequency drift so the signal may be understood.

Controlling RF Gain

Radio Frequency (RF) GAIN (volume control) amplifies all received signals, including interference. If the RF gain is too high then received signals will vary in loudness and could be distorted because of interference. To ensure a constant level of received signal, the RF GAIN should be used in conjunction with AGC. When AGC is used, RF GAIN should be at maximum.

Using Automatic Gain Control (AGC)

AGC will maintain a constant level of received sound. When AGC is not used, the RF GAIN must be adjusted.

Using the 2182 kHz Instant Selector

Pressing the 2182 pad will (usually) instantly select 2182 kHz, full power, H3E.

Testing the Radiotelephone Alarm Generator

Always a two-finger operation. Pressing two clearly marked pads (eg ALARM & TEST) at the same time will actuate the Radiotelephone Alarm without transmission.

Using the Radiotelephone Alarm Generator

This is always a two-finger operation. Pressing two clearly marked pads (eg ALARM & SEND) at the same time will actuate the Radiotelephone Alarm.

Ideally the Alarm should be sent for 1 minute, however, most transmitters will actually send the Alarm for 45 seconds.

ANTENNAS

Isolators are manually operated levers or switches which will connect antennas to different transmitters, or to earth.

VHF Whip Antennas are usually 1 or 2 metres in length, about the diameter of a broomstick, mounted vertically, as high as possible.

MF/HF Whip Antennas may be of different lengths eg from 2 to 5 metres, and can be 'top-loaded' to accommodate a range of frequencies. They are normally mounted on the bridge wings.

MF/HF Wire Antennas are lengths of copper wire running fore/aft, or around the superstructure. All insulators should be kept clean to prevent leakage currents running to earth over encrusted salt and dirt.

Construction of an Emergency MF Antenna
This is applicable to all types of craft including yachts.

Construction of an emergency antenna may be achieved by using a long length of wire (the longer the better) or co-axial cable. The wire should be stretched to its full length and fixed to insulators at either end. Aerial insulators may be made of ceramic, glass, or plastic but if none are to hand, a short length of nylon cord (or even a twisted plastic bag!) will suffice, tied to either end of the wire and made fast to eg a mast halyard at one end and the gunwale or rail, at the other.

The wire antenna should be mounted as high and as close to vertical as possible with a connection taken from the antenna to the Antenna Tuning Unit (ATU). This connection (or *lead-in*) should be short in order to avoid electrical losses. If using co-axial cable (the type of cable which connects your TV aerial to the set), strip back the shielding braid so that only the copper conductor makes contact with the ATU socket.

TUNE for the highest reading possible on the ATU.

The ATU is earthed direct to the hull, or to an earthing plate in the hull in the case of a wooden or fibreglass vessel.

If using the back-stay of a yacht as your antenna, reverse-loop the ATU lead-in on to the back-stay . This will help to prevent corrosion from water which would otherwise run down and into the connection. *(See Appendix 4 titled: AMSA Safety Education Article number 64 – Care, Maintenance and Installation of HF Marine Radio Transceivers – Technical Notes).*

BATTERIES

Batteries hold an electrical charge, contain a highly corrosive and dangerous electrolyte and under certain conditions, for example if the discharge terminals were short-circuited or if hydrogen gas were ignited by a cigarette, could explode.

On board ship, batteries may be in every day use supplying low voltages to radio communications equipment. However, whether the batteries are in use or are designated an emergency supply only, regular maintenance is imperative so that an emergency supply voltage can always be relied upon.

Usually there will be 2 groups or 'banks' of batteries so that while one bank is discharging (supplying voltage to equipment), the other bank may be charged.

Batteries must be kept in a well ventilated, dry place, normally on an upper deck of the ship.

Different Kinds of Batteries and their Characteristics
The batteries could be lead-acid type (similar to car batteries) and could also be of the Nickel-Cadmium (or NICAD) type.

Lead Acid Batteries contain a number of cells, each cell having a discharge voltage of about 2 volts. Each cell contains a number of lead peroxide (positive) plates and lead (negative) plates, in a solution (electrolyte) of dilute sulphuric acid. Chemical action between the plates produces electricity which is taken as discharge voltage. Hydrogen gas is given off during discharge and smoking is forbidden near batteries to guard against the risk of explosion. A group of 6 – 2 volt cells connected together in one casing gives us a 12 volt battery.

Voltage when fully discharged will fall to about 1.8 volts per cell.

Nickel-Cadmium batteries (NICADs) are more expensive and more durable than lead-acid cells. The electrolyte is Potassium Hydroxide and is highly caustic. Plates are Nickel Hydroxide (Positive) and Cadmium (Negative).

Fully charged discharge voltage is about 1.4 volts per cell falling to 1.3 volts when under load. Specific gravity of the electrolyte is around 1.220 but as this does not change, hydrometer readings will give us no indication of the cell's condition. Therefore the best indication of the cell's state of charge is the on-load voltage reading. When this drops to about 1.0 volt the cell should be charged.

Charging

Battery connections when charging, from the charging source to the battery, are from positive to positive and negative to negative. When current is forced through the cell in a reverse direction to the discharge, chemical action within the cell is also reversed, so that the cell is charged.

Constant Current Charge

The charge source must be several times greater than the potential of the battery to be charged eg about 230 volts for a 12 volt battery. A large resistor is used to limit the charging current which varies only slightly as the battery charges. With this type of charging there is a danger of excessive gassing in the final stages and a large heat loss through the resistor.

Constant Voltage Charge

The charging source may be up to 25% greater than the potential of the battery to be charged eg 8 volts for a 6 volt battery, 15 volts for a 12 volt battery. A constant voltage is applied and current is high to begin with but decreases as the battery charges. There is a danger of plate-buckling in the early stages of this method, because of excessive current.

Trickle Charge

This method uses a small current to keep the battery in a constant state of charge. The charging rate may be as little as 1% of the battery Ampere Hour rating.

A low charge would be about 10% of the A.H. rating.

A high charge would be about 50% of the A.H. rating.

A battery charger will be part of the radio communications equipment.

Connecting Batteries

Batteries may be connected in one of 3 ways, to form a 'bank' of cells, these methods of connection are:-

1) **Series** connection, when positive terminals are connected to negative terminals. In series connection, the voltages add up arithmetically ie. two 12 volt batteries in series give an output of 24 volts.

2) **Parallel** connection, when positive is connected to positive, and negative to negative. Batteries in parallel give an output voltage equal to the voltage of one battery, but with an extended period of discharge eg two 12 volt batteries in parallel will give an output of 12 volts, but for twice as long as the discharge period of one battery. That is, the combined capacity of the batteries – measured in Ampere Hours (AH) – is double the capacity of a single battery.

3) **Series-Parallel** is a means of combining the two above methods so that for example, two batteries could be connected in parallel, then connected in series to another two batteries which are also connected to each other in parallel. Voltage is consistent with the diagram.

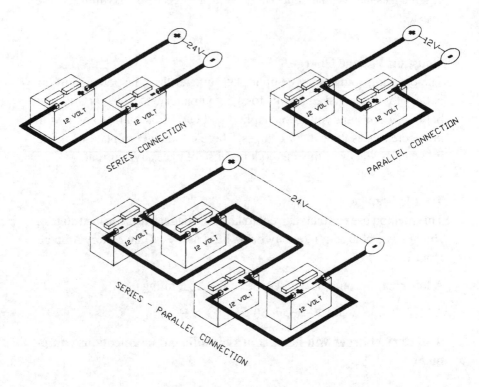

Ampere-Hours (AH) is the Capacity of a battery and is determined by the size of the plates. A battery with a capacity of 70 AH will supply 7 Amps for 10 Hours, or say 70 Amps for 1 hour. Two batteries of 70 AH when connected in parallel have a combined AH capacity of 140 AH. The capacity of a battery gives an indication of the uses to which it may be put.

Maintenance of Batteries

Batteries must be kept clean. Condensation, dirt, and spilled acid, will allow leakage currents to conduct. Wash any corrosion from the lead terminals with water, then coat terminals with petroleum jelly to guard against more corrosion forming. Keep connections tight and clean.

Always keep electrolyte at the correct level (about 1 cm above plates) and use distilled water for topping up. Tap water may be used if absolutely necessary but this will introduce impurities into the cell which ultimately will shorten its life.

Take 'on load' voltage readings regularly. These are readings taken when the transmitter is in use and for a 24 volt battery should be about 22.8 volts. Charge regularly. Regulations specify weekly readings to be entered in battery log.

Specific Gravity (SG) is the weight of the electrolyte (or any substance in fact) compared to the weight of an equal volume of water.

Because the chemical composition of the electrolyte changes during charge/discharge cycles, the specific gravity also changes. When the specific gravity is low, internal resistance of the cell is high and the converse applies ie. a high specific gravity indicates a low internal cell resistance and the possibility of higher current being drawn from the cell.

Taking regular **Specific Gravity** readings is the best way to discover the state of lead acid batteries. This is done with a **hydrometer**.

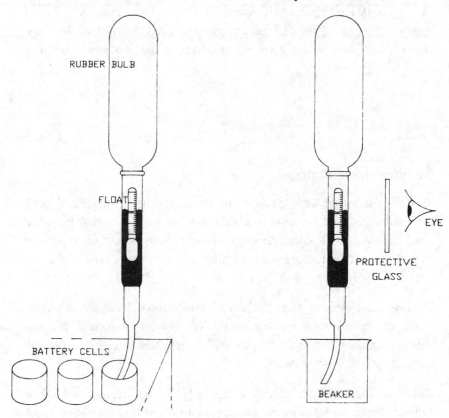

The rubber bulb is depressed and the nozzle inserted into the cell electrolyte, then the rubber bulb released. Electrolyte is drawn up into the glass barrel causing the float to rise. The specific gravity reading is taken where the acid level meets the graduated float.

SG readings should be between 1250-1280. If the readings are below 1250, put the battery on charge.

Regulations specify readings to be taken monthly and entered into the battery log.

Safety

Do **not** smoke near batteries. An explosion could result.

Keep cell tops on to prevent dirt from falling into the electrolyte and so that acid droplets will not spray out.

Loosen tops when charging to allow free discharge (and thereby prevent build-up) of hydrogen gas.

Wear protective goggles or spectacles to prevent eye damage when servicing batteries.

Wear protective rubber gloves.

Immediately wash off any splashed acid with eyewash or water.

Potassium Hydroxide from NICAD cells should be washed off with boracic acid solution.

Metal cases of NICAD cells are *LIVE*.

Be careful not to take metal objects eg spanners, aluminium jugs, into battery rooms. An explosion could result if these were set down across terminals.

Uninterrupted Power Supplies (UPS) are necessary to ensure an uninterrupted supply voltage to specific equipment (MF/HF transceiver and Ship Earth Station) in the event of mains power supply failure. If the ship's main supply is interrupted for more than 2 milliseconds, the UPS will switch the equipment to either standby generator, or to batteries, for emergency supply voltage.

Terminology:

Power Supplies
When we speak of power supplies to electrical equipment, we are speaking of voltage, measured in volts, symbol **V** or **v**.

Supply voltage could be from the ship's mains, from the ship's auxiliary (standby) engine room supply, or from batteries.

Current is measured in Amperes or Amps, symbol **I**.

Power is measured in Watts, symbol **W**, and is given by **V** multiplied by **I**.

Resistance is measured in Ohms, symbol **R**.

Current is drawn from the voltage supply by equipment which offers a physical or ohmic resistance to current flow. The greater the Resistance the lower the current flow will be.

If we now look at Ohm's Law, then **I = V/R**.

If **R** decreases then **I** will increase and the supply voltage will not change.

The resistance of a circuit could decrease because of a fault in the equipment, which would result in more current being drawn from the supply than the equipment is designed for.

Fuses are used to protect the equipment, and the operator, from such occurrences. The type of fuse normally used in radio communications equipment is a small glass tube with aluminium end caps and the actual fuse wire enclosed within.

Ceramic fuses are often used in switchboards and in heavy duty power supply equipment, as are Circuit Breakers, which will interrupt the power supply in the event of a fault and can be reset when the fault has been repaired.

Each fuse has a Rating. This is the maximum current the fuse will carry indefinitely. When the rating is exceeded, the fuse 'blows', which then prevents voltage and current from flowing in the circuit.

Current is the dangerous component of electricity so it is of especial interest to us. Even a relatively small current of say 30 milliamps (occurring, for example, because of carelessness when handling batteries) could be fatal.

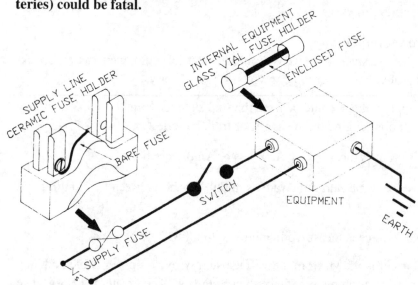

Remember to switch off the supply before replacing any fuses.

SURVIVAL CRAFT RADIO EQUIPMENT

Survival Craft Station
A mobile station in the maritime mobile service or the aeronautical mobile service intended solely for survival purposes and located on any lifeboat, life-raft or other survival equipment.

Portable Two-Way VHF Radiotelephone Apparatus must be able to transmit and receive on 156.8 MHz (Channel 16) and at least one other frequency.

UHF Radiotelephony equipment is not compulsory but if it is used, must be able to transmit and receive on 457.525 MHz.

Search And Rescue Radar Transponder (SART)
At least one Radar Transponder to be carried on every ship of 300 GRT and above and two Radar Transponders on ships of 500 GRT and above.

Radar Transponders may be carried in Survival Craft and may be incorporated in a VHF DSC EPIRB.

After 'switch on', The SART remains in 'Standby' mode (for up to 96 hours) until activated by a 9 GHz radar emission which then causes the SART to transmit a signal appearing as a straight line of 'blips' on the radar display.

Emergency Position Indicating Radio Beacon (EPIRB)
A 'float-free' EPIRB which can be located in the COSPAS-SARSAT system must be carried on board. The EPIRB will transmit locatable and identifiable signals for up to 48 hours after activation.

Digital Selective Calling (DSC)

DSC is a method of automatically alerting ship or shore stations, individually or collectively, by dialling a 9-digit numerical code (MMSI station ID – see para 2.2) into the DSC controller, which is permanently connected to the ship's main transceiver. The DSC controller enables digitally coded signals to be transmitted on specific frequencies. A special keypad allows for automatic transmission of Distress Alert and other keypads give access to selections from various menus.

CALL FORMAT SPECIFIER

This enables selection of **Call Type**, **Address**, **Priority,** etc, by use of keypads, so that different types of call may be programmed into the Controller prior to transmission.

The words Call Format Specifier do not actually appear on the DSC Controller.

The Call Format Specifer is a function of the controller which gives the operator access to the following Selections, which are displayed on a small VDU, to be chosen as required by simply pressing keypads.

Distress Call
Distress calls may be addressed to all ships or to individual stations. Other information such as working (Radiotelephony) frequencies and type of distress, may also be included.

All Ships Call
Calls may be addressed to all ships for routine matters in much the same way as the Distress Call situation above.

Call to Individual Station is achieved by inserting the ID (MMSI number) of the required station into the requisite memory selection of the Controller.

Geographic Area Call
A specific geographic area may be defined on some Controllers, and calls addressed to that area.

Group Call
A group of stations may be called by insertion of group identity MMSI.

Automatic/Semi-Automatic Service

DSC Controllers have a facility which will enable telephone numbers to be inserted into the call. After contacting the required coast station, the Controller accesses the public telephone network of the country concerned through equipment in the coast station. the operator on board then picks up the Radiotelephone handset on the MF/HF or VHF transceiver and speaks to the called telephone subscriber ashore. This may be accomplished either automatically, or semi-automatically via a coast station telephone operator.

CALL ADDRESS SELECTION WITH THE MMSI (MARITIME MOBILE SERVICE IDENTITIES) NUMBER SYSTEM

The Nationality Identification

MMSI identification numbers are allotted to each vessel. The MMSI consists of 9 digits the first 3 of which indicate the country of ship registration, as defined by the ITU Maritime Identification Digits or MID.

Maritime Identification Digits or MID.

Each country has been allotted 3 figures eg 503 for Australia, 232 for the United Kingdom of Great Britain and Northern Ireland, 563 for Singapore. This is the MIDs. The remaining 6 figures of the MMSI are allocated by the administration concerned.

Group Calling Numbers are intended for use by stations calling simultaneously more than one ship. The first figure is zero.

Coast Station Numbers

Coast station IDs begin with two zeros eg 005030331 Perthradio

MMSI Number With Three Trailing Zeros

These IDs are for use by vessels which can reasonably expect to use automatic access devices for public switched networks world-wide. Ship stations which require automatic access on national or regional level should be assigned 1 trailing zero or 2 trailing zeros. MMSIs without trailing zeros should be assigned to all other vessels requiring numerical ID.

CALL CATEGORIZATION

This is a function of the Controller designed to give a range of priorities which may be selected by the use of keypads.

The words Call Categorization do not appear on the Controller.

Distress
Note that on receipt of a distress call options for **Relay** or **Acknowledgement** will appear.

Note also that transmitted relays and acknowledgments are received by all ships within range and not just by the vessel in distress.

Note further that acknowledgement of DSC distress alert by ships should be by radiotelephony in the same band on which the alert was received. For example, if the alert was received on 2187.5 kHz then acknowledgement should be on 2182 kHz. If the alert was received on 4207.5 kHz then the acknowledgement should be on 4125 kHz.

Urgency
Used when the Distress signal is not justified, to broadcast a message concerning the safety of a ship, aircraft, other vehicle or a person.

Safety
Indicates that a message concerning safety of navigation or an important weather message is to follow.

Ship Business
To be conducted on frequencies for Routine use.

Routine
Used for automatic & semi automatic radiotelephone calls, ship business, etc.

CALL TELECOMMAND AND TRAFFIC INFORMATION

This function of the Controller provides for selection of the type of call we wish to initiate. The words Call Telecommand and Traffic Information do not appear on the Controller.

Distress Alerts may be transmitted quickly by pressing one keypad marked typically, with **Dist**. or **Distress**. After pressing for 5 seconds a Distress Alert will be broadcast to all stations. This will always contain the Identity of the calling vessel, its position and the time of the position, if the Controller is interfaced with the ship's navigation system.

If the Controller is not interfaced with the ship's navigation system, the position must be updated at least every 12 hours. A position more than 12 hours old will not be accompanied by a time. This omission serves as advice to recipients that the position is unreliable.

When programming a distress **call**, options which appear are:-

On MF, H3E or J3E.

On VHF, F3E/G3E, Simplex.

On MF/VHF, distress relay, distress acknowledgment.

On MF/VHF it may be possible to select the type of distress eg collision, cargo shift etc, Otherwise the indicator **Undesignated Distress** will appear.

If position of the Distress is not available, the digit **9** should be inserted ten times, and if time of Distress is unavailable, the digit **8** inserted ten times.

Other Calls

All other MF/HF ship-shore calls should not be on the DSC Distress Alerting frequencies but on the national frequencies allocated to the coast station. The ship DSC transmit frequency of **2189.5 kHz** coupled with the shore transmit frequency of **2177 kHz** have been allocated internationally for **Routine** calls. Other calls could be Routine calls for ships business, or telephone calls.

On MF or VHF, the DSC controller may be programmed to call a coast station with a subscriber telephone number and initiate the setting up of an automatic or semi-automatic telephone call.

The frequency on which a DSC call was received will be indicated.

On MF/HF the selections **H3E**, **J3E**, or **end of call** appear.
On VHF, **F3E/G3E simplex** or **duplex** or **end of call**.

The signal **RQ** is usually selected at **End Of** Sequence (**EOS**) to request mandatory acknowledgment, which then includes the signal **BQ**.

Medical transport calls should be given the category **Urgency** and be addressed to **all stations**.

Test calls may be initiated on MF/HF but not on VHF and usually by prior arrangement with a coast station.

Unable to Comply indicates to the calling station that the called station cannot comply (cooperate) immediately with a request (due for example, to called ship operator busy with ship operations). If the called ship's operator is **Able to Comply** sometime later, then a call should be initiated to the calling vessel indicating a working frequency or channel.

Working Frequency Information
Where a working frequency/frequencies or channel is indicated in the call but is not acceptable to the called station, the called station should immediately reply with a proposal for an alternative frequency or channel.

General Principles of Narrow Band Direct Printing (NBDP) and Telex Over Radio (TOR) systems. Ability to Use Maritime NBDP and TOR Equipment in Practice

*Note: **Narrow Band Direct Printing (NBDP)** telegraphy is a means of communication whereby a printed or hard' copy is available at each communications terminal. NBDP is also known as radiotelex, and is used for NAVTEX broadcasts.*

NARROW BAND DIRECT PRINTING (NBDP)

Ships at sea may communicate with shore-based telex machines by calling coast radio stations on frequencies specifically allocated for NBDP use. NAVTEX is a system utilising NBDP in the shore-ship direction only, for the broadcast of, for example, MSI.

NBDP Systems

Automatic Systems
Whenever possible automatic procedures should be used i.e. the calling subscriber should contact the called subscriber directly without the aid of an operator.

a) In ship to shore working, the ship station calls the coast station on a predetermined coast station receive frequency, using the direct-printing equipment and the identification signal of the coast station assigned in accordance with Appendix 38, or the coast station identity assigned in accordance with Appendix 43.

b) The coast station's direct-printing equipment detects the call and the coast station responds directly on the corresponding coast station transmit frequency, either automatically or under manual control.

For example, the message is typed into memory, address and coast station code inserted into call procedure, calling frequency specified and by simply pressing a key the message is sent automatically for delivery via International/National telex networks.

Semi-Automatic Systems
The telex operator of the international exchange of the land station

country selects the called subscriber directly if automatic procedures or single-operator procedures cannot be applied. ie after typing message into memory, a coast station operator is called (on telex), who will then forward the message for delivery via International/National telex networks.

Manual Systems

The land station operator applies manual procedures if automatic, single-operator or semi-automatic procedures are not possible, i.e. the coast station operator is called and the message typed direct, not having been saved in memory.

*For radiotelex communication there are two modes in common usage, Automatic Request Query (**ARQ**) and Forward Error Correction (**FEC**).*

ARQ is the method commonly used for ship-shore-ship communication. The message is prepared on a keyboard, then transformed to an international 7-Unit code (like Morse code) before transmission.

During quiet periods an 'idle' signal is transmitted when it is possible, for example, to access radiotelex equipment at coast radio stations. The station which initiates contact becomes the Master, while the called station becomes the Slave. This situation prevails for the duration of the circuit no matter which station is sending. Direction of traffic flow is changed when the characters +? are transmitted.

On receipt of control signals (called **CS1** and **CS2**) the Master station will begin transmitting its information in blocks of 3 characters. Blocks will be numbered alternately '**1**' or '**2**', dependent on reception of CS1 or CS2 Control Signals. Each block is transmitted in 210 milliseconds after which a pause of 240 milliseconds allows for reception of CS1/CS2 Control Signals in the event of request for repetition.

After each correctly received block, the receive station transmits a Control Signal CS1 or CS2 of 70 milliseconds duration, followed by a pause of 380 milliseconds. Timing is controlled by the Master station clock, thus each block uses a standard time cycle of 450 milliseconds so that the time interval between end of received signal and beginning of transmit signal will remain constant. This time will be varied if a mutilated character is received. When signals are continuously mutilated and time between blocks exceeds 15 seconds, the equipment will revert to 'Standby'. This will also occur when the Master station transmits an 'End Of Work' signal.

To sum up, information is transmitted in blocks of 3 characters for printed (hard copy) reception. The equipment transceiver repeatedly requests 'bits' of information until the correct transmit/receive code has been established. This operation is a function of the equipment and, in itself, need not concern the operator. However if the connection is poor, there is a greater possibility of the circuit 'dropping out'. Care should be taken therefore when deciding which frequency to use.

FEC is a stream of information in one direction only ie a broadcast. It may be used by coast stations for the transmission of eg meteorological information to ship-board equipment which has been left in receive mode.

Phasing signals, which will be recognised by the receiver as the start of a broadcast, begin each transmission. Information is sent in blocks of 3 units or characters and each block is sent twice. After transmission of the first character, a group of 4 characters is transmitted, then the first character is re-transmitted. The receiving equipment will accept the above sequence only when reception of the second transmission is seen to duplicate the first. Errors in the received copy will print out as a space or an asterisk. FEC is used in NAVTEX, where a received error rate of more than 4% will not print out.

By using Selective FEC (**SELFEC** or **SFEC** on the equipment Mode Selection) messages may be transmitted to individual ships in this way by prior arrangement (when a ship is in port for example) and coast stations will expect a receipt to be given at some later time.

Vessels carrying HF NBDP (Radiotelex) and travelling exclusively in areas covered by these services, are exempted from the carriage of INMARSAT EGC receivers (for the reception of MSI).

ISS/IRS Arrangement
The transmitting station is known as **INFORMATION SENDING STATION (ISS)** and the receiving station as the **INFORMATION RECEIVING STATION (IRS)**. Whether using paired frequencies or a single frequency timing is arranged so that transmission takes place at any given time, in one direction only.

Master and Slave
When working direct to a coast station using ARQ the station which called first is the Master (ISS), while the called station is the Slave (IRS).

The signal **+?** must be sent at the end of each transmission. The signal **+?** is the equivalent of 'Over' in Radiotelephony and hands control to the receive station which then becomes the transmit station.

At end of work the Master station must close down the circuit by pressing KKKK.

Radio Telex Number
Every radio telex machine has an identification number.

Ship station selective call numbers are comprised of 5 digits and coast station selcall numbers are 4 digits followed by the additive sign + eg **0331+** for Perthradio.

Answerback
Every telex machine has an answerback code. This is usually comprised of letters identifying country, followed by letters identifying a business, then the identification number of the machine. EG AA indicates Australia, RADCOM indicates an abbreviated business name, 123456 is the machine identity. The answerback would then appear as RADCOM AA123456.

Answerback of a ship's machine could be the country code followed by the identification number by itself or combined with letters. It is imperative that an answerback is printed at the end of every telex or radiotelex communication. The answerback is proof of communication.

Numbering of the SSFC (Sequential Single-Frequency Code)
Messages typed onto a keyboard are firstly converted into the 5-Unit Baudot Code which is then itself converted into a 7-Unit code for transmission. The ISS then transmits the information in blocks of 3 characters (letters, figures, or spaces). The IRS converts the received information from 7-Unit back to Baudot for printing. After reception of each block of 3 characters, the IRS sends a Control Signal back to the ISS which then transmits the next block. Any mutilated characters will print out as an asterisk (or possibly a space) at the IRS. Transmitted blocks may be repeated up to 32 times, to ensure error-free reception. This is the method used for ARQ.

In FEC each block is transmitted twice and if the second transmission does not exactly duplicate the first, then an asterisk (or possibly a space) will print out at the IRS.

TOR (TELEX OVER RADIO) EQUIPMENT

The operation of TOR equipment (an earlier version of HF Radiotelex) has been described above.

Controls and Indicators

POWER ON/OFF BUTTON & LAMP
ALARM OFF BUTTON
SELECT BUTTONS AND LAMPS FOR :- STANDBY – ARQ –
    FEC – SELFEC -CLEAR – OVER (+?)
LAMPS INDICATING:- STORE FULL – SENDING OR RE-
    CEIVING TRAFFIC -ERROR – PHASE – ELEMENTS A – Z
    (1 – 0) CONTROL SIGNALS CS1-CS2 STATUS- -MASTER –
    SLAVE
FUNCTION SWITCH FOR TEST – CW – TUNE – ETC
TUNE METER & LAMP (FOR TRANSMITTER TUNING)

Keyboard Operation
Information is typed directly onto the keyboard to be either stored in memory or prepared on tape for transmission.. Message start is always **ZCZC** and message end always **NNNN**.

Typing direct to IRS is only possible when SEND or TX lamp is on. The characters +? must be typed at the end of each transmission (or the OVER button pressed) so that IRS can reply.

Knowledge of the Usage of INMARSAT SYSTEMS. Ability to Use INMARSAT Equipment or Simulator in Practice.

INMARSAT-A SHIP EARTH STATION

Satellite Acquisition

Ships may select the satellite which covers the Ocean Region in which they are sailing by manually typing instructions for antenna alignment onto the keyboard, or by selecting AUTO TRACK for the antenna to automatically self-align.

Telex Services may be accessed by selecting Telex, then appropriate CES. International country dialling codes are available in the INMARSAT Users Handbook. Some CESs provide a Telex Group Call Service by which a message can be transmitted shore to ship to a group of ships eg ships of the same registry, or ships of the same fleet. Each INMARSAT-A SES has a unique INMARSAT identification number.

The following reporting systems may be accessed via INMARSAT:
Weather reporting;
Position reporting;
AMVER;
AUSREP;
JASREP;
SISCONTRAM;
Pollution reporting;
Locust reporting; and
Quarantine reporting.

Telephone Services may be accessed by selecting Telephone, then appropriate CES. See the INMARSAT User's Handbook for services provided.

Data and Facsimile Communications

Some ships are equipped with computers which can interface with the satellite terminal to provide ship-shore-ship data transmission and reception. Facsimile may be received on board, with the correct modem interface, whilst delivery by facsimile ship-shore is identical to the process for Telephony.

INMARSAT EGC RECEIVER

Pre-programming an SES for EGC Message Reception

EGC messages are broadcast over an entire Ocean Region but will be accepted only by EGC receivers logged in to that Ocean Region and which are in the geographical area addressed. It is therefore important that the ship's position be regularly updated.

EGC messages may be broadcast to circular or rectangular areas or to selected groups of vessels whether grouped by fleet, flag, or geographical area.

Message types may be selected for reception or rejection. For example if a vessel does not have DECCA radio navigation equipment then DECCA messages may be de-selected.

Selecting Operating Mode for EGC Reception

INMARSAT-C terminals incorporate the EGC receiver which can be accessed via pull-down menus to select/de-select message types.

On INMARSAT-A terminals the EGC receiver is not normally incorporated into the equipment. It is more likely to be a stand-alone receiver and may have its own printer. Follow the manufacturer's instructions for EGC reception.

Note that <u>EGC Only</u> will permit reception of EGC messages *<u>ONLY!</u>*

See Admiralty List of Radio Signals Volume 3 for INMARSAT coverage of World Wide Navigational Warning System.

INMARSAT-C SHIP EARTH STATION

Components of an INMARSAT-C Terminal

Each **SES** consists of a DTE (Data Terminal Equipment) and a DCE (Data Circuit Terminating Equipment). Messages are prepared and formatted in the DTE for transmission by the DCE to satellite. In the reverse direction, the DCE receives messages for display/printing/storage at the DTE.

The SES comprises:-
Small omni-directional antena DCE

Electronics Unit	**DCE incorporating Word Processor, Buffer Memory, and EGC receiver)**
Keyboard	**DTE**
Monitor	**DTE**
Printer	**DTE**

Entering/Updating Position

This is easily done manually by the use of pull-down menus or automatically if the SES is interfaced with the ship's navigational system. Position updating should be done at least every 4 hours.

Usage of an INMARSAT-C Terminal

Each SES has a unique INMARSAT Identification number. Pull-down menus enable easy usage of all facets of these terminals. The SES must be 'logged-in' to the satellite covering the region which the ship is traversing. This can be done manually, or automatically by instructing the SES to 'log-in' to the strongest signal. However care should be taken if this is done as there will be times when signals from different satellites will be of the same received strength.This results in the SES 'searching' from one signal to the other so that SES to Satellite contact could be lost from time to time. Some EGC messages may not be received if this occurrs.

Sending and Receiving Text Messages

Messages may be typed directly into the word processor and either transmitted directly from the screen to a selected address, or filed into Memory for transmission at a later time. Received messages may be printed, saved in memory or on disk, or both.

Fault Locating

Proficiency is required in Elementary Fault Localisation by means of Built-in Measuring Instruments or Software in Accordance with the Equipment Manuals and Elementary Fault Repair such as Replacement of Fuses, Indicator lamps, and the like.

Modern equipment is usually capable of self-diagnostic tests, which will indicate to the operator which part of the equipment is faulty. Other indications of faults may be lamps which illuminate when a fault is detected by the equipment, fuses indicating they are 'blown' or circuit breakers which have 'tripped', meters giving incorrect readings, and visual displays on built-in screens, or printed read-outs.

Faults commonly occur in power supply circuits when fuses blow, or connections become loose, for example, because of vibration.

Fault-finding is the domain of skilled technicians and can only be accomplished by actual practice on 'live' equipment. Remember that each piece of equipment has a manual which usually includes basic fault-finding advice, flow-charts, fuse location, etc.

READ THE MANUAL!

Part 3

Operational Procedures and Detailed Practical Operation of GMDSS System and Subsystems

Global Maritime Distress and Safety System (GMDSS)

GMDSS

GMDSS is an enhanced **S**earch **A**nd **R**escue **(SAR)** system using automation and advanced satellite and HF/MF/VHF radio communication technology.

The System is a combination of:-

Satellite communications;

MF/HF/VHF Radiotelephony;

Radiotelex (Direct Printing Telegraphy);

Survival Craft Radar Transponders (SARTs);

Satellite EPIRBs;

Data transmission; and

Digital Selective Calling (DSC).

Some companies are manufacturing the full range of GMDSS equipment to be accessed from one keyboard and one VDU (screen).

GMDSS provides for the rapid alerting of SAR authorities and vessels and the efficient deployment of coordinated SAR services.

Rescue Coordination Centres (RCCs) and Maritime Rescue Coordination Centres (MRCCs) are responsible for the promotion of efficient organisation of SAR services and for coordinating the conduct of SAR within a SAR region.

Australian Maritime Safety Authority (AMSA)
Some countries have well equipped maritime rescue organisations in place and augment their services by use of a ship reporting system. A good example of such an organisation is the **Australian Maritime Safety Authority (AMSA)** which operates the Australian Ship Reporting System **(AUSREP)** participation in which is compulsory for all ships within a designated region.

Shipboard Communications Equipment required for the GMDSS will be specified by geographical operating areas and not, as at present, by ship size. This will give ships the option of fitting out for either satellite or HF/MF/VHF radio systems, or a combination of both.

Maritime Safety Information (MSI)

GMDSS also provides for the transmission of navigational and weather information (Maritime Safety Information or MSI), using Satellite services and Radio in the Medium Frequency (MF), High Frequency (HF), and Very High Frequency (VHF) Bands.

SEA AREAS AND THE GMDSS MASTER PLAN

Sea Area A1

Means an area within the radiotelephone coverage of at least one VHF coast station in which continuous DSC alerting is available, as may be defined by a country's Government.

Sea Area A2

Means an area, excluding sea area A1, within the radiotelephone coverage of at least one MF coast station in which continuous DSC alerting is available, as may be defined by a country's Government.

Sea Area A3

Means an area, excluding sea areas A1 and A2, within coverage of an INMARSAT geostationary satellite in which continuous alerting is available.

Sea Area A4

Means an area outside sea areas A1, A2, and 3.

The GMDSS Master Plan (to be in place on 1st February 1999) Includes:-

the list of VHF coast stations for sea area A1;
the list of MF coast stations for sea area A2;
the list of HF coast stations for sea areas A3 and A4;
the list of INMARSAT CESs;
the list of COSPAS-SARSAT MCCs;
operational and planned NAVTEX services;
distress message routing in the INMARSAT/RCC ship-shore distress
    alerting networks;

SESs commissioned for RCC operations;

the list of proposed and operational SafetyNET services;

maritime SAR regions, RCCs and associated shore-based facilities; and information on shore-based facilities in the GMDSS.

Contracting Governments must provide the necessary shore-based facilities for the following satellite and terrestrial radiocommunication services:-

a) INMARSAT;
b) COSPAS-SARSAT;
c) The Maritime Mobile Service in the bands between 156 and 174 MHz (VHF);
d) The Maritime Mobile Service in the bands between 4,MHz and 27.5 MHz (HF); and
e) The Maritime Mobile Service in the bands between 415 kHz and 535 kHz and between 1.605 MHz and 4 MkHz (MF).

The Master Plan together with the Radio Regulations is the basis for determining GMDSS equipment requirements for ships.

WATCHKEEPING ON DISTRESS FREQUENCIES

Whilst at sea ships shall maintain a Digital Selective Call (DSC) watch on the DSC Distress and Safety frequencies in the bands in which they are operating.

DSC watch must also be kept on 2187.5 kHz.

Ships having no Satellite Earth Station must keep continuous watch on 2187.5 kHz, 8414.5 kHz and one other frequency in the HF Band.

Watch must also be maintained on DSC Channel 70 VHF, unless exempted by the government concerned.

In the Guard Band 2173.5 – 2190.5 kHz a listening watch must be maintained on 2182 kHz.

In the Guard Band 156.7625 – 156.8375 MHz a listening watch must be kept on 156.8 MHz (Channel 16).

An Automatic Watch for reception of Maritime Safety Information (MSI) broadcast by INMARSAT must also be kept.

Coast stations with watchkeeping responsibilities will maintain a continuous Automatic Watch for Distress Alerts from ships at sea and, where so equipped, a DSC watch also.

FUNCTIONAL REQUIREMENTS OF SHIP STATIONS

Current regulations require that every ship (unless exempted) shall be capable of performing the following system requirements in an efficient manner:-

Of transmitting ship-to-shore DISTRESS alerts by at least two separate and independent means, each using a different radiocommunication service;

Of receiving shore-to-ship DISTRESS alerts;

Of transmitting and receiving ship-to-ship DISTRESS alerts;

Of transmitting and receiving Search And Rescue coordinating communications;

Of transmitting and receiving locating signals;

Of transmitting and receiving Maritime Safety Information;

Of transmitting and receiving general radiocommunications relating to the management and operation of the vessel;

Of transmitting and receiving bridge-to-bridge communications; and

Of transmitting DISTRESS alerts by carriage of either:
a)   406 MHz (COSPAS-SARSAT) EPIRB; or
b)   'L' Band (INMARSAT) EPIRB.

Operational Requirements
The SOLAS Convention requires that equipment performing the functions detailed above shall be simple to operate and, wherever appropriate, be designed for unattended operation.

CARRIAGE REQUIREMENTS OF SHIP STATIONS

Minimum Requirements for all ships
VHF equipment with Radiotelephony on Channels 16,13,6, and DSC on Channel 70; A satellite EPIRB or a VHF EPIRB;
A SART (which may be incorporated in a VHF DSC EPIRB);

A NAVTEX receiver if sailing in areas in which a NAVTEX service is provided;

An INMARSAT EGC receiver if sailing in areas in which a NAVTEX service and MSI by HF NBDP is not provided;

A 2182 kHz watchkeeping receiver (not required after 1999); and

A Radiotelephone Alarm Signal (Two-Tone Alarm) generator (not required after 1999).

Carriage requirements for GMDSS ships may be summarized as follows:-

Sea Area A1

VHF equipment and either a VHF EPIRB or a satellite EPIRB.

Sea Area A2

VHF and MF equipment and a satellite EPIRB.

Sea Area A3

VHF, MF, a satellite EPIRB and either HF or satellite communications equipment.

Sea Area A4

VHF, MF and HF equipment and a satellite EPIRB; and,

All ships will carry equipment for receiving MSI broadcasts.

Small craft (eg. pleasure craft, fishing vessels etc) in the Australian region should equip with a SSB HF transceiver with a maximum power of at least 150 Watts, and carry an EPIRB.

Documents to be carried on board

1. Station Radio Licence.
2. GMDSS Certificate/s.
3. Radio Log.
4. Alphabetical List of Call-signs and/or Numerical Table of Identities of Stations Used by the Maritime Mobile Service and Maritime Mobile Satellite Service.
5. List of Coast Stations and Land Earth Stations in the GMDSS.
6. List of Ship Stations.
7. Manual for Use by the Maritime Mobile and Maritime Mobile-Satellite Services.

NOTE: Governments may, under appropriate circumstances (for example, when ships are sailing only within range of VHF Coast Stations) exempt ships from the carriage of the documents mentioned in paragraphs 4 to 7 above.

SOURCES OF ENERGY FOR SHIP STATIONS

Three separate power supply sources must be available for radiocommunications equipment: These could be:

1) ship's main supply;

2) ship's auxiliary supply; or

3) batteries.

MEANS OF ENSURING AVAILABILITY OF SHIP STATION EQUIPMENT

The SOLAS Convention requires that each country shall ensure that the radio equipment is maintained to provide the availability of those functions detailed above

Ships trading in areas A1 and A2, shall ensure this availability by using at least one of the following methods:-

Duplication of equipment (Multi-fitting);

Shore-based maintenance; and

On-board maintenance.

On ships trading in areas A3 and A4, this availability shall be ensured by using a combination of at least two of the following methods:-

Duplication of Equipment
In order to satisfy the Convention requirements where there is no on-board maintainer vessels will be required to carry duplicate equipment as follows:

Area A1   DSC VHF Radio;

Area A2   DSC VHF Radio and either MF OR HF Radio, or SES;

Area A3   DSC VHF Radio and either HF Radio or SES; and

Areas A3 & A4 DSC VHF Radio, HF Radio.

Separate antennas are required for duplicate radio installations.

LICENCES, RADIO SAFETY CERTIFICATES, INSPECTIONS AND SURVEYS

Operator's Certificates

Categories of certificate for GMDSS equipped ship stations are:

a) First Class Radio Electronic Certificate;

b) Second Class Radio Electronic Certificate; and

c) GMDSS General Operator's Certificate

Australian requirements are that at least two GMDSS General Operator's Certificates must be carried on GMDSS vessels (ie vessels subject to the SOLAS Convention) where an on-board maintainer is not carried. On these vessels, duplication of certain equipment is necessary.

Licences

Every transmitting station must hold a licence or radio safety certificate issued by the government of the country to which the station is subject. The licence must be kept on board and is subject to renewal following survey and inspection by a qualified official of the administration concerned.

Categories of Ship Stations

a) Category 1 stations maintain a continuous service.

b) Category 2 stations maintain a service for 16 hours each day.

c) Category 3 stations maintain a service for 8 hours each day.

d) Category 4 stations maintain a service which is either of shorter duration than that of Category 3 stations, or of a duration which is not fixed by the Regulations.

INMARSAT Usage in the GMDSS

INMARSAT-A SHIP EARTH STATION

Distress Communications

BEFORE INITIATING DISTRESS PROCEDURE:-

a) Consider your geographical location and consider which Coast Earth Station is best situated to help;

b) If time permits, write down your DISTRESS message before transmission;

c) Be aware of User's Handbook instructions for your equipment, particularly with regard to DISTRESS calling; and

d) The Master's authority is required for DISTRESS, URGENCY, or SAFETY calls.

Remember that misuse of the DISTRESS facility may result in the de-commissioning of your SES.

Use of the Distress Facility

On INMARSAT-A or INMARSAT-B equipment the DISTRESS alert is usually activated by either:-

a) Pressing a prominent DISTRESS key; or

b) Dialling sequential numbers eg. 999.

Doing this will provide telephone or telex access to an MRCC or CES and, if necessary, will pre-empt a Channel for DISTRESS communications. However with some equipments a Routine call may need to be initiated to an MRCC or CES for initial Distress alerting.

In addition to DISTRESS button, or sequential number dialling, INMARSAT offers the following options for DISTRESS calling:

a) Priority 3, providing a range of services for the seafarer in DISTRESS. Any call prefixed with Priority 3 is given automatic connection to an RCC;

b) Routine priority, to make a connection with an MRCC;

c) Routine priority to make a connection with an RCC operating a land based Ship Earth Station, (eg. Falmouth, U.K.); and

d) Routine priority to communicate with any organisation or person, whether ashore or at sea.

If a DISTRESS call is mistakenly directed to a Coast Earth Station (CES) not listed in the appropriate Table (see INMARSAT User's Manual) the DISTRESS call will be intercepted by dedicated CESs in each Ocean Region.

Some RCCs may be accessed only by the use of Routine priority calling and dialling. Operators should be aware of requirements for RCC access in the Ocean Region which they are traversing.

Access to Rescue Services
Note: Where connection to a Maritime Rescue Co-ordination Centre is switched via a public telephone or telex network, some Coast Earth Stations offer "GA+" (meaning Go Ahead) in telex mode or "Proceed to select" in telephone mode.

The Ship Earth Station is then required to dial in the access code for the desired MRCC.

This method of access is called a 'DIVERSION' in the INMARSAT User's Handbook.

Satellite Acquisition
Any Distress indicator is recognised at the NCS or CES and a satellite channel instantly assigned. If all channels are occupied, then a channel will be pre-empted.

Telex and Telephony Distress Calls
Ship to MRCC (Telex Mode)

a) Ensure that NBDP or TELEX mode is selected.

b) Appropriate CES is selected.

c) Initiate the call in accordance with the Ship Earth Station manufacturer's instructions for a DISTRESS call.

d) Check that correct Coast Earth Station ID is received.

e) Where DIVERSION is offered and chosen, GA+ must be answered with the correct sequence of digits.

f) Check correct answer-back is received from the point of connection. If no answer-back is received in 15 seconds, then repeat the call.

If calling in DIVERSION and no answer-back is received in 40-90 seconds, or the codes NA, NCH, NP, ABC, CI, DER, INF, JEE, or a wrong answer-back are received, then repeat the call.

g) Transmit the DISTRESS message.

h) Keep line open as far as is practicable to receive an acknowledgment, and further instructions from the responsible authorities.

Ship to MRCC (Telephone Mode)

a) Ensure that TELEPHONE mode is selected.

b) Appropriate Coast Earth Station is selected.

c) Initiate the call in accordance with the Ship Earth Station manufacturer's instructions for a DISTRESS call.

d) Only where DIVERSION is offered, receive the 'Proceed to select' tone of 1.5 seconds duration.

Only where DIVERSION is chosen, immediately key in the correct sequence of digits.

e) Receive 'Ringing' answer from the point of connection. If any other tone or no tone is heard, then hang up and try again.

f) Receive a voice confirmation from the point of access.

g) Transmit the DISTRESS message, using DISTRESS procedure.

h) Receive acknowledgment and instructions from the responsible authorities.

Procedures for Distress Calls should follow International Distress procedures for Radiotelephony. However, no Regulation prevents the use by a mobile station or a mobile earth station in Distress of any means at its disposal to attract attention, make known its position, and obtain help.

Rescue Co-ordination Centres Associated with the Coast Earth Stations

Each Coast Earth Station (CES) is linked to a Maritime Rescue Co-ordination Centre (MRCC) or Rescue Coordination Centre (RCC), which is responsible for promoting the efficient co-ordination of SAR.

INMARSAT-C SHIP EARTH STATION

Distress and Safety Services
The same services are available via INMARSAT-C SES as are available via INMARSAT-A and INMARSAT-B terminals, except that no voice communication is possible.

Sending a Distress Alert
This is easily done by either using the Distress menu selection, or pressing Distress pads. For example, by pressing two keypads on the electronic control unit, at the same time, for 5 seconds. Don't forget the initial Alert will contain information which may require updating of position, course, and time.

Sending a Distress Priority Message
Distress is selected on the pull-down menu and information easily inserted into the message eg position, nature of distress, course, speed, CES selection.

The INMARSAT-C Safety Services
Enhanced Group Call (EGC) is included as an integral part of INMARSAT-C SES stations. MSI (via the INMARSAT SafetyNET service) is received and may be printed, saved on disk, or both. Ship position reports may be automatically transmitted at set times and some CES have a facility enabling shore companies to take information (eg data polling) reports from ships.

2-Digit Code Safety Services
See the INMARSAT-C Maritime Users Manual for tables of 2-digit codes, which give immediate access through selected CES, to a number of services including Safety.

2-Digit codes for Perth CES for example are 38 for Medical Assistance,
39 for SAR,
42 for MSI.

INMARSAT ENHANCED GROUP CALL (EGC)

Purpose of the EGC System
Information Providers such as Meteorological Bureaux, RCCs, Shipping Companies, etc, broadcast messages to individual ships, or groups of

ships, whether grouped by flag, fleet, or geographical area, using the SafetyNet or FleetNet systems.

SafetyNet is for the broadcast of Maritime Safety Information (MSI) to be received on EGC receivers. The SES must be logged-in to the Ocean Area for which MSI is required and the ships position regularly updated at least ever 4 hours, to ensure reception of messages relevant to the ship's position. *(See Admiralty List of Radio Signals Volume 3 for INMARSAT coverage of World Wide Navigational Warning System).*

Use of the **FleetNET** system enables Information Providers such as Shipping Companies to broadcast messages to individual ships or groups of ships. The NCS will download Network Identification Codes for reception of EGC messages to, for example, a fleet of ships. These codes are known as CNID (Closed Network Identification for EGC) and DNID (Closed Network Identification for Data), and are NOT inserted by the SES operator.

All Ships Messages and INMARSAT System Messages
All EGC receivers will accept messages addressed to All Ships and INMARSAT System messages.

Classes of INMARSAT-C SES and their EGC Reception:-

Class 0, Option 1 SES for standalone EGC reception only;

Class 0, Option 2 SES is a standalone EGC receiver added on to an INMARSAT-A terminal;

Class 1, SES cannot receive EGC messages; and

Class 2, SES can receive EGC messages when not engaged in normal (non-EGC) message transmission and reception.

NAVTEX

THE NAVTEX SYSTEM

International NAVTEX Service is the coordinated broadcast and automatic reception on the frequency 518 kHz of Maritime Safety Information by means of narrow-band direct-printing telegraphy using the English language.

National NAVTEX Service is the broadcast and automatic reception of Maritime Safety Information by means of narrow-band direct-printing telegraphy using frequencies other than 518 kHz and languages as decided by the countries concerned.

Purpose of NAVTEX
NAVTEX is a low-cost, International System, for the automatic dissemination of MSI by narrow band direct printing, to ships. NAVTEX provides shipping with MSI and urgent information by automatic print-out from a dedicated receiver. It is suitable for use in all sizes and types of vessel.

NAVTEX Frequencies
490 kHz – 518 kHz (the main frequency used) – 578 kHz – 4209.5 kHz

Reception Range
With a range of about 400 nm, each station has a maximum transmission power of 200-1000 watts which is regulated so that interference between stations is minimised. As stations serve prescribed areas it is desirable that ship's personnel closely monitor received messages in the broadcast zones they are transiting.

Message Format (Transmitter ID, Message Type, Message Number)
ZCZC (phasing signal) always appears as the first group on the first line of the received message. This is followed by 4 letters and figures designated as follows:-

Station ID   single letter (B1);

Message type   single letter (B2); and

Message number   2 figures (B3-B4).

NAVTEX RECEIVER

This equipment has few controls and is easy to operate. Transmitter Areas and Message Types can be easily selected and changed.

Selection of Transmitters is accomplished by choosing a single letter of the alphabet which designates a particular transmitter. Transmitters in each NAVTEX broadcast area are allocated identification letters alphabetically from A to Z ranging clockwise around the coast.

Selecting Message Type and Messages which cannot be Rejected

Prefix Codes

At the beginning of each message is a sequence of subject indicator characters B1, B2, B3 or B4, which enables the NAVTEX receiver to identify different types of message. Receiver settings may be used to reject messages which have already been received or which are of no interest to a vessel eg. DECCA messages for ships without DECCA equipment.

The **B1** character identifies the transmitter. Transmitting stations are identified by a single letter of the alphabet.

The **B2** character identifies the subject in compliance with the following list.

A **Navigational warnings (see Note 1).**
B **Meteorological warnings (see Note 1).**
C **Ice reports.**
D **Search And Rescue information (see Note 1).**
E **Meteorological forecasts.**
F **Pilot service messages.**
G **DECCA messages.**
H **LORAN messages.**
I **OMEGA messages.**
J **SATNAV messages.**
K **Other electronic navaid messages (see Note 2).**
L **Navigational warnings – additional to letter A (see Note 3).**
V **Special services.**
W **Allocated by.**
X **NAVTEX.**
Y **Panel.**
Z **No messages on hand.**

B3 and **B4** indicate the NAVTEX message serial number.

71

NOTES:-

1) Cannot be rejected by the receiver.

2) Messages concerning radio-navigation services.

3) Should not be rejected by the receiver (continuation of B2 subject group "A").

Message Format

This is vitally important as any Format error on the part of transmit stations could result in message loss. Message content and priority is strictly controlled to avoid duplication and to ensure delivery of messages. Priority is determined by message originators.

PRIORITIES:-

VITAL

For immediate broadcast.

IMPORTANT

For broadcast at the next available period when the frequency is unused.

ROUTINE

For broadcast at the next scheduled transmission.

DISTRESS

The NAVTEX broadcast is not suitable for distress traffic. Only the initial distress message should be re-transmitted on NAVTEX.

In which case:-

B1 = Transmitter identification;

B2 = D (Distress);

B3 = 0; and

B4 = 0.

Operation

At least 8 hours before sailing switch on the NAVTEX receiver to ensure all Maritime Safety Information is received.

Use of Subsidiary Controls and Changing Paper

Other controls typically available on NAVTEX receivers are DIMMER (controls display brightness), FEED (extends paper feed), ALARM (disables/enables audible alarm), TEST (equipment self-tests) and POWER (On/Off).

Emergency Position Indicating Radio Beacons (EPIRBs)

SATELLITE EPIRBS

Ship's Masters should ensure that all crew members know the type of EPIRB and SART carried and the way in which the equipment operates. In particular the line provided on EPIRBs is for tying the EPIRB to a liferaft or lifeboat and NOT for tying the EPIRB to its housing or to its float-free mechanism. The line on a SART is for lashing the equipment to as high a point as possible in a survival craft, and NOT for towing the SART in the water!

Basic Characteristics of Operation on 406 MHz

The COSPAS-SARSAT System uses 4 satellites at a height of about 1000 km, in near-polar orbits, and operated jointly by the USSR (COSPAS) and an American/Canadian/French consortium (SARSAT). Together the satellites scan every part of the earth's surface every 2-3 hours and receive signals from EPIRBs on 406 MHz and 121.5 MHz.

Local User Terminal (LUT)

In over 20 countries there are a number of ground stations known as Local User Terminals or LUTs, which track the satellites and receive information from them as they pass overhead.

EPIRB signals received at the satellite exhibit a frequency shift known as Doppler Effect. This information is relayed to the LUT where it is computer analysed and related to the satellite position, which is always precisely known. The EPIRB position can then be accurately determined.

Signals on **406 MHz** are radiated with a power of 5 watts, and processed at the satellite. Doppler shift is measured and position information transferred in digital form to the LUT, when both EPIRB and LUT appear in the satellite footprint at the same time. The satellite also stores this information for subsequent downloading to LUTs. Therefore, even if a LUT is not visible to the satellite when it receives an EPIRB transmission on 406 MHz, the EPIRB location will still be passed to the next LUT which is passed over by the satellite and to all other LUTs. On 406 MHz one satellite pass should be sufficient for position fixing and accurate to within 5 nm.

Basic Characteristics of Operation on 1.6 GHz

EPIRBs operating in the 1.6 GHz range with a radiated power of 1 watt utilise the INMARSAT system. The EPIRB is connected through a modem to the ship's navigational system and is continually updated with the vessel's position. The EPIRB must be float-free and may also be manually activated when the ship's last position and ID is transmitted. This type of EPIRB formerly known as the L-Band EPIRB (because it operates over a range of frequencies in the 1.6 GHz range known as the L-Band) is now known as the **INMARSAT-E** EPIRB.

121.5 MHz Including Homing Functions

When the satellite observes an EPIRB and a LUT within its 'footprint' simultaneously, signals will be relayed immediately to the LUT for analysis. The detection range for 121.5 MHz EPIRBs extends as a circular area (of Satellite Visibility) to a radius of 1620 nm (3000 km) from the LUT. The low powered (75 milliwatts) emission on 121.5 MHz is also used as a 'homing' beacon by aircraft in SAR operations. Specialised direction finding equipment enables the aircraft to search large areas and eventually locate survivors. If the satellite cannot 'see' a LUT when it sees an EPIRB, then signals on 121.5 MHz will not be relayed. EPIRB positions determined on these frequencies should be accurate to within 20 nm but because of the low powered transmission more than one satellite pass may be necessary.

There is a Local User Terminal at Alice Springs and another at Wellington, New Zealand, with a resultant combined area of Satellite Visibility embracing Australasia. Both of these LUTs are controlled by Mission Control Centre in Canberra, Australia.

Unlike the INMARSAT system, which offers a full range of telecommunication facilities, the COSPAS-SARSAT system is limited to the position fixing of EPIRBs. The Australian Maritime Safety Authority (AMSA) coordinates SAR activities and is linked to the LUTs at Alice Springs and Wellington.

Information Contents of a Distress Alert

Transmissions will include, besides position information, a unique ID number which identifies the ship and provides other details to INMARSAT or COSPAS-SARSAT. In some cases the type of Distress may be manually programmed into the EPIRB prior to transmission.

Manual Usage

EPIRBs should be mounted clear of the ship's superstructure so that they are easily accessible for maintenance, so that they may be carried into a life raft and manually activated. A line is provided to tie the EPIRB to the liferaft or lifeboat before placing it overboard.

Some EPIRBs may be activated by a switch operated from the bridge where a ship is in Distress but in no immediate danger of sinking.

Float-free Function

EPIRBs in the GMDSS must be float-free, ie fitted with a hydrostatic release mechanism which activates under pressure if immersed to a depth of 4 metres. The EPIRB should be mounted clear of superstructure, lines, stays, etc, so that it may float free of its housing in the event of immersion and automatically begin transmitting.

Routine Maintenance

EPIRBs should be kept clean, so that the manual switch does not become clogged with salt or dirt and visual checks made to ensure the EPIRB will not foul on stays, or lines in the event of immersion.

Testing

EPIRBs have a function switch enabling testing to be carried out without actual transmissions taking place. When in TEST position there will be a visual indication for example a flashing strobe light.

Checking Battery Expiry Date

EPIRBs are powered by batteries which have a 10 year 'shelf-life' and which should be replaced according to the manufacturer's instructions. A plate indicating when the battery should be replaced is affixed to the EPIRB by the manufacturer. This should be regularly checked. When in continuous use the batteries will last for 50 hours only.

Cleaning of the Float-free Mechanism

Cleaning should be part of the routine maintenance programme.

Satellite Compatible EPIRBs

121.5/243 MHz EPIRBs, which have been manufactured after 1990 and which meet the Australian Spectrum Frequency Agency standard 'SMA 241', emit a signal whose 'shape' is suitable for position fixing within the COSPAS-SARSAT system. Positions may be determined to within 20 nm.

VHF DSC-EPIRB

The Main Technical Characteristics

VHF DSC EPIRBs transmit Distress Alerts on the VHF DSC Channel 70. These EPIRBs must also incorporate a SART operating in the 9 GHz band. The following have previously been covered on pages 74 to 75:-

Information Contents of a Distress Alert;
Manual Operation;
Float-free Function;
Routine Maintenance;
Checking Battery Expiry Date; and
Cleaning of the Float-free Mechanism.

Search and Rescue Radar Transponder (SART)

THE MAIN TECHNICAL CHARACTERISTICS

The SART transmits a signal over a swept band in the 9 GHz range (X Band) which appears as a straight line of 'blips' on the radar screen.

Operation
SARTs may be packed in a life raft, incorporated in a float-free EPIRB or mounted in a float-free housing eg. above a ship's bridge. The SART may be switched on manually or automatically if immersed in water (eg. when incorporated in a VHF DSC EPIRB).

After switch-on the SART will remain in standby mode (for up to 96 hours) until a radar transmission in the 9 GHz range, from ship or aircraft, triggers a response. When this occurs, transmission from the SART appears as a perfectly straight line of twelve 'blips' (radiating outward from the SART position) on the radar screen. Distance between the 'blips' is 0.6 nm. As a searching vessel draws closer to the SART position the blips will increase in size to form arcs and then complete circles, when the vessel is close enough to trigger the SART transmission continually. The appearance of arcs or circles on the radar screen should serve as a warning to **SLOW DOWN!** This distinctive signal will be an obvious indication to Search And Rescue craft. When transmitting the SART also 'rings' thereby indicating to the survivors that searchers are close.

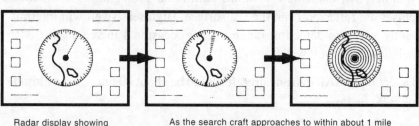

Radar display showing	As the search craft approaches to within about 1 mile
the SART 12-dot blip	of the SART the 12 dots will change to wide arcs than
code. (Bearing 30°)	into complete circles as the SART is closed.

Some SARTs have a second facility called ANTI-COLLISION mode. When this mode is used, the SART will transmit 5 pulses ranging over 1 nautical mile.

A cord is provided to lash the SART to eg. the upper part of an inflatable raft doorway.

Range of a SART Transmitter

Range is dependent on a a number of factors including: the height at which the SART is mounted in a liferaft or lifeboat (as high as possible outside the liferaft canopy – inside the canopy will severely restrict the range); the type of radar; the height of its antenna on the SAR vessel; and the range at which the radar is set.

A radar set to short range will reduce brightness of the 'blips' so they may not be seen at extreme range. Radars set to longer ranges, eg. 12 nm or 24 nm, emit longer pulses which are of of higher energy than at short ranges. So a SART transmission should be clearly identifiable on both the 24 nm or the 12 nm ranges and all 12 'blips' will be seen.

Having identified the SART transmission, the radar range may be progressively shortened as the SAR vessel approaches but on shorter ranges all 12 'blips' will not be seen.

The IMO performance standard for SARTs recommends a mounted height of 1 metre above sea level to give a range of 5 nm, given a ship's interrogating radar antenna mounted at 15 metres above sea level. A well mounted SART may have a detection range of up to 10 nm in good weather.

Skill of operation of the SAR vessel radar is also a consideration where the ability to eliminate 'clutter' may affect SART detection. Flat calm weather will contribute to multi-path radar propagation when radar pulses are reflected from the sea surface. High seas/swell will intermittently vary the detection range due to occasional increases in SART height above sea level.

Detection of SARTs from aircraft should be about 40 nm with an altitude of 3000 metres.

Routine Maintenance of a SART

SARTs should be kept clean, so that the manual switch does not become clogged with salt or dirt.

Checking Battery Expiry Date

A plate indicating when the battery should be replaced is affixed to the SART by the manufacturer. This should be regularly checked.

Distress, Urgency and Safety Communication Procedures in the GMDSS

DISTRESS COMMUNICATIONS

The Distress call shall have absolute priority over all other transmissions. All stations which hear it shall immediately cease any transmission capable of interfering with the distress traffic and shall continue to listen on the frequency used for the emission of the distress call. This call shall not be addressed to a particular station and acknowledgment of receipt shall not be given before the distress message which follows it is sent. The Distress call and message shall be sent only on the authority of the master or person responsible for the ship, aircraft or other vehicle carrying the mobile station or ship earth station.

DSC Distress Alert

The Distress Alert is a digital selective call using a distress call format in bands used for terrestrial radio communication.

The Definition of a Distress Alert is the rapid and successful reporting of a Distress incident to a unit which can provide co-ordinate assistance.

The transmission of a Distress Alert requires the master's authority and indicates that a mobile unit or a person is in distress and requires immediate assistance.

Distress Alerts fall into two categories:-

1) SHIP-SHORE Alerting via satellite from Ship Earth Stations or EPIRBs; and

2) SHIP-SHIP and SHIP-SHORE Alerting via DSC in the MF, HF and VHF Bands.

Transmission of a DSC Distress Alert may be made by simply pressing the Distress keypad or by selection from menus on the DSC Controller where other information, eg type of Distress and subsequent means of communication (Radiotelephony or NBDP), may be entered.

Alerts are normally received by other vessels in the vicinity and by Coast Radio stations or Coast Earth Station which then relay communications to and from Rescue Co-ordination Centres (RCCs) or Maritime Rescue Co-ordination Centres (MRCCs).

Transmission of a Shore to Ship Distress Alert Relay

A station or a RCC which receives a Distress Alert shall initiate the transmission of a shore-to-ship Distress Alert Relay addressed, as appropriate, to all ships, to a selected group of ships or to a specific ship by satellite and/or terrestrial means. DSC controllers store received Distress (and other) calls and have the function of Distress Relay, accessed by 'soft key' menus.

Transmission of a Distress Alert by a Station Not Itself in Distress

A station in the mobile or mobile-satellite service which learns that a mobile unit is in distress shall initiate and transmit a Distress Alert in any of the following cases:

a) when the mobile unit in Distress is not itself in a position to transmit the Distress Alert; and

b) when the master or person responsible for the mobile unit not in Distress or the person responsible for the land station considers that further help is necessary.

Receipt and Acknowledgment of a DSC Distress Alert

Remember that coast stations acknowledge by DSC and ship stations acknowledge by radiotelephony in the first instance. Ship stations may acknowledge by DSC if there is no acknowledgment from a coast station within 5 minutes.

Acknowledgment Procedure by Radiotelephony

Ship stations should acknowledge DSC Distress Alerts by Radiotelephony in the band in which the DSC Alert was received, using Radiotelephony Distress procedures:-

MAYDAY

NAME, RADIO CALLSIGN OR MMSI IDENTITY OF STATION SENDING DISTRESS MESSAGE (SPOKEN 3 TIMES)

THIS IS (OR DELTA ECHO IN CASE OF LANGUAGE DIFFICULTIES)

NAME, RADIO CALLSIGN OR MMSI IDENTITY OF STATION ACKNOWLEDGING (SPOKEN 3 TIMES)

RECEIVED MAYDAY *(OR ROMEO ROMEO ROMEO MAYDAY)*

Acknowledgment Procedure by NBDP

MAYDAY

NAME, RADIO CALLSIGN OR MMSI OF STATION SENDING DISTRESS ALERT

THIS IS or *DE*

NAME, RADIO CALLSIGN OR MMSI OF STATION ACKNOWLEDGING

RRR MAYDAY

Receipt and Acknowledgment by a Coast Station

Coast stations and appropriate CESs in receipt of Distress Alerts shall ensure that they are routed as soon as possible to a RCC. Receipt of a Distress Alert is to be acknowledged as soon as possible by a coast station or by a RCC, via a coast station, or an appropriate CES. A coast station using DSC to acknowledge a Distress call shall transmit the acknowledgment on the Distress calling frequency on which the call was received and should address it to all ships. The acknowledgment shall include the MMSI of the ship whose Distress call is being acknowledged.

Receipt and Acknowledgment by a Ship Station

Ship or SESs in receipt of a Distress Alert shall, as soon as possible, inform the master or person responsible for the ship of the contents of the Distress Alert.

In areas where reliable communications with one or more coast stations are practicable ship stations in receipt of a Distress Alert should defer acknowledgment for a short interval, so that receipt may be acknowledged by a coast station.

Ship stations operating in areas where reliable communications with a coast station are not practicable which receive a Distress Alert from a ship station which is, beyond doubt, in their vicinity shall, as soon as possible and if appropriately equipped, acknowledge receipt and inform a RCC through a coast station or CES.

However a ship station receiving an HF Distress Alert shall not acknowledge it but shall set watch on the Radiotelephone Distress and Safety traffic frequency associated with the Distress and Safety calling frequency on which the Distress Alert was received.

If acknowledgment by Radiotelephony of the Distress Alert received on the MF or VHF Distress alerting frequency is unsuccessful, acknowledge receipt of the Distress Alert by responding with a DSC call on the appropriate frequency.

HANDLING OF DISTRESS ALERTS

Preparations for Handling of Distress Traffic
On receipt of a DSC Distress Alert ship stations and coast stations shall set watch on the radiotelephone Distress and Safety traffic frequency associated with the Distress and Safety calling frequency associated with the Distress Alert frequency. Coast stations and ship stations should additionally set watch on the NBDP frequency associated with the Distress Alert signal if it indicates that subsequent working is to by NBDP.

Distress Traffic Terminology

MAYDAY is the Distress Signal

SILENCE MAYDAY (PRONOUNCED SEELONCE M'AIDER) used to impose silence on stations causing interference to Distress traffic

PRUDONCE indicates that complete silence is no longer necessary on a Distress frequency

SILENCE FINIS (PRONOUNCED SEELONCE FEENEE) to indicate cessation of Distress traffic

Testing DSC Distress and Safety Calls
Means should be provided for the testing of DSC equipment without radiation. Testing on the exclusive DSC Distress and Safety calling frequencies should be avoided as far as possible by using other methods. There should be no test transmissions on the DSC calling channel on VHF. However when testing on the exclusive DSC Distress and Safety calling frequencies on MF and HF is unavoidable, it should be indicated that these are test transmissions. The test call should be acknowledged by a coast station with normally no further communication between the two stations.

On-Scene Communications normally take place on the MF and VHF bands on frequencies designated for Distress and Safety traffic by Radiotelephony or NBDP. These communications, between the ship in Distress and assisting units, relate to the provision of assistance to the ship or the rescue of survivors. When aircraft are involved in on-scene communications they are normally able to use 3023 kHz, 4125 kHz, and 5680 kHz. In addition, SAR aircraft can be provided with equipment to communicate on 2182 kHz or 156.8 MHz, or both, as well as on other maritime mobile frequencies.

Search and Rescue Operation

SAR coordinating communications are for the co-ordination of ships and aircraft participating in a SAR operation following a Distress Alert. They include communications between RCCs and any 'on-scene commander' (OSC) or 'coordinator surface search' (CSS) in the area of the Distress incident. An efficient communications network should exist between RCCs and between RCCs and their associated CESs and coast stations. SAR action in response to any Distress situation will be achieved through co-operation among SAR administrations able to provide assistance. The first RCC is the RCC associated with the shore station which first acknowledged the alert and assumes responsibility for co-ordinating Distress traffic and controlling the SAR operation. The first RCC may delegate these responsibilities to another station.

URGENCY AND SAFETY COMMUNICATIONS

The Meaning of Urgency and Safety Communications

These have priority over all other communications except for Distress, and include:-

navigational and meteorological warnings and urgent information (MSI);

ship-to-ship safety of navigation communications;

ship reporting communications;

support communications for SAR;

other urgency and safety messages;

communications relating to navigation, movements and needs of ships;

OBS (weather) messages for an official meteorological service.

Procedures for DSC Urgency and Safety Calls

DSC, on the Distress and Safety calling frequencies, should be used to advise shipping of the impending transmission of Urgency and vital navigational and Safety messages, except when the transmissions take place at routine times. Where the subsequent transmission of an Urgency and vital navigational or Safety message takes place on a working frequency, the call should indicate the frequency which will be used.

Urgency Communications

The Urgency Signal consists of the words PAN PAN. In Radiotelephony each word of the group shall be pronounced as the French word 'panne'. The Urgency call format and the Urgency Signal indicate that the calling station has a very urgent message to transmit concerning the safety of a mobile unit or a person.

In Radiotelephony the Urgency message shall be preceded by the Urgency Signal repeated 3 times and the ID of the transmitting station.

In NBDP the Urgency message shall be preceded by the Urgency Signal and the ID of the transmitting station. The Urgency call format or Urgency Signal shall be sent only on the authority of the master or the person responsible for the mobile unit carrying the mobile station or mobile earth station. The Urgency call format or the Urgency Signal may be transmitted by a land station or a CES with the approval of the responsible authority. When an Urgency message which calls for action by the stations receiving the message has been transmitted the station responsible for its transmission shall cancel it as soon as it knows that action is no longer necessary.

Error correction techniques shall be used for Urgency messages by NBDP. All messages shall be preceded by at least one carriage return, a line feed signal, a letter shift signal, and the Urgency signal PAN PAN.

Urgency communications by NBDP should normally be established in FEC mode. The ARQ mode may subsequently be used when it is advantageous to do so.

Medical Transports

This term refers to any means of transportation by land, water or air, whether military or civilian, permanent or temporary, assigned exclusively to medical transportation. When control is under a competent authority of a party to a conflict, or of neutral States and of other States not parties to an armed conflict, when these ships, craft and aircraft assist the wounded, the sick and the shipwrecked.

For the purpose of announcing and identifying medical transports the urgency signal shall be followed by the addition of the single word MEDICAL in NBDP and by the addition of the single word MAY-DEE-CALL pronounced as in French 'medical' in Radiotelephony.

A Medical Transport message should contain:-

ID of the medical transport;

position of the medical transport;

number and type of vehicles in the medical transport; and

intended route.

Safety Communications

In a terrestrial system the announcement of the Safety message shall be made on one or more of the Distress and Safety frequencies using DSC. A separate announcement need not be made if the message is to be transmitted through the maritime mobile-satellite service. The Safety Signal and message shall normally be transmitted on one or more of the Distress and Safety traffic frequencies or via the maritime mobile-satellite service or on other frequencies used for this purpose. The Safety Signal consists of the word SECURITE. In Radiotelephony, it shall be pronounced as in French.

The Safety call format or the Safety Signal indicates that the calling station has an important navigational or meteorological warning to transmit. In Radiotelephony, the safety message shall be preceded by the safety signal and the ID of the transmitting station.

Error correction techniques shall be used for safety messages by NBDP. All messages shall be preceded by at least one carriage return, a line feed signal, a letter shift signal and the safety signal SECURITE.

Safety communications by NBDP should normally be established in the broadcast (FEC) mode. The ARQ mode may subsequently be used when it is advantageous to do so.

COMMUNICATION BY RADIOTELEPHONY WITH STATIONS OF THE OLD DISTRESS AND SAFETY SYSTEM

All of this section is still applicable in the GMDSS.

Radiotelephone Alarm Signal

Known as the Two Tone Alarm, the radiotelephone alarm signal is two alternating tones of 1300 Hz and 2200 Hz the duration of each tone being 250 milliseconds. The Two Tone Alarm should be sent before the Distress call if time permits and for at least 30 seconds but not more than one minute. When sent by a coast radio station it shall be followed by a long dash of 10 seconds.

The purpose of the Two Tone Alarm is to attract the attention of watchkeepers and to activate 2182 kHz watchkeeping receivers when transmitted on 2182 kHz. The Two Tone Alarm shall only be used to announce:-

1) **that a Distress call or message is about to follow;**

2) **the transmission of an urgent cyclone warning which should be preceded by the Safety signal. This procedure may only be used by authorized coast radio stations; and**

3) **the loss of a person or persons overboard, when assistance of other vessels is required and cannot be obtained by use of the Urgency signal alone.**

The Distress Signal

Consists of the word MAYDAY pronounced as the French expression m'aider. All Distress traffic by radiotelephony should be preceded by the word MAYDAY, which indicates that a ship, aircraft or other vehicle is threatened by grave and imminent danger and requests immediate assistance. The Distress signal has absolute priority over all other transmissions.

Radiotelephony Distress procedure consists of:-

1) the RadiotelephoneTwo Tone Alarm Signal (time permitting);

2) Distress call; and

3) Distress message.

Note that all SOLAS communications are in SIMPLEX mode which requires use of the word 'OVER'.

Distress Call
> *MAYDAY MAYDAY MAYDAY*
> *THIS IS*
> *SHIP'S NAME SHIP'S NAME SHIP'S NAME*

Distress Message
> *MAYDAY*
> *SHIP'S NAME* – (or other identification)
> *POSITION*
> *NATURE OF DISTRESS AND TYPE OF ASSISTANCE*
> *REQUIRED*
> *OTHER INFORMATION WHICH MAY FACILITATE*
> *RESCUE*

Watchkeepers hearing a Distress call should immediately prepare to copy the Distress message, which should follow, into the radio log or ship's log using UTC.

Acknowledgement of Distress Message
Acknowledgement (which does not require the master's permission) should take the following form

> *MAYDAY MAYDAY MAYDAY*

(MAYDAY should be the first word in all Distress calls)

> *SHIP'S NAME SHIP'S NAME SHIP'S NAME*
> (of vessel in Distress)

> *THIS IS*

> *SHIP'S NAME SHIP'S NAME SHIP'S NAME*

> *ROMEO ROMEO ROMEO MAYDAY* or *RECEIVED MAYDAY*

Vessel's acknowledging Distress must send a message containing:- name; position; speed; and ETA at the Distress position.

Silence may be imposed by the station in control of the Distress SAR on stations which are causing interference by use of the words *SILENCE M'AIDER (*pronounced *SEELONCE MAYDAY*). Other stations may impose silence on interfering stations by use of the words *SEELONCE DISTRESS*.

No transmissions are permitted whilst the Distress is in progress, except for Distress transmissions. Normal radio working may be resumed when the controlling station broadcasts *SILENCE FINI* (pronounced *SEELONCE FEENEE*) as follows:-

> *MAYDAY*
>
> *ALL STATIONS ALL STATIONS ALL STATIONS*
>
> *THIS IS STATION NAME* (once only)
>
> *TIME*
>
> *NAME AND CALL SIGN* (or other ID) *OF VESSEL IN DISTRESS*
>
> *SEELONCE FEENEE*

Distress Traffic Terminology

Distress traffic consists of all messages relating to the immediate assistance required by the mobile station in Distress.

MAYDAY is the radiotelephone Distress signal and indicates that a ship, aircraft or other vehicle is threatened by grave and imminent danger and requests immediate assistance.

SEELONCE MAYDAY pronounced as the French expression "Silence M'aider" may be used by the station in Distress or the station controlling Distress traffic, to impose radio silence on stations which are causing interference.

SEELONCE DISTRESS pronounced as the French expression "Silence" followed by the word 'DISTRESS" may be used by other stations which consider it essential to impose silence on interfering stations.

PRUDONCE pronounced as the French word "Prudence" may be used by the station controlling Distress traffic to indicate that restricted working may be resumed, as complete silence is no longer necessary on a frequency being used for Distress traffic.

MAYDAY RELAY pronounced as the French expression "M'aider relais" is used by a station,which though not itself in Distress, is about to re-broadcast a Distress message on behalf of a station which is in Distress.

SEELONCE FEENEE pronounced as the French expression "Silence finis" is broadcast by the station controlling Distress traffic, and indicates that Distress traffic has ceased on a frequency which has been used for Distress, and normal working may be resumed. SEELONCE FEENEE is broadcast as follows:-

MAYDAY
HELLO ALL STATIONS (3 TIMES)
THIS IS
NAME/ID OF STATION SENDING THE MESSAGE
TIME OF HANDING-IN OF THE MESSAGE IN UTC
NAME & CALLSIGN OR MMSI OF STATION WHICH WAS
   IN DISTRESS SEELONCE FEENEE

Transmission of a Distress Message by a Station Not Itself in Distress Mayday Relay

This procedure uses the expression M'AIDER RELAIS pronounced MAYDAY RELAY, and shall be transmitted under the following conditions:-

1) **when the station in Distress is unable to transmit the Distress message;**

2) **when the master of the vessel or other vehicle not in Distress or the person responsible for the land station, considers that further help is necessary; and**

3) **when although unable to render assistance, it has heard a Distress message which has not been acknowledged.**

The Distress message should be transmitted exactly as received, and shall always be preceded by the call below:-

MAYDAY RELAY MAYDAY RELAY MAYDAY RELAY
THIS IS
SHIP'S NAME SHIP'S NAME SHIP'S NAME (or other ID)

MAYDAY
RECEIVED THE FOLLOWING FROM
(NAME OF VESSEL IN DISTRESS AND TIME CALL WAS
RECEIVED IN UTC) (Distress message exactly as received).

Vessels using this procedure must indicate that they themselves are not in Distress and shall take all steps to notify authorities who may be able to render assistance.

Note that vessels receiving MAYDAY RELAY broadcast by a coast radio station, shall not acknowledge receipt unless the master of the vessel has confirmed that the ship concerned is able to assist.

Medical Advice may be obtained from most coast stations by use of either a radiotelegram addressed to 'RADIOMEDICAL' (followed by the name of the coast station) or by requesting a direct link call to a rostered medical adviser. Use of the Urgency signal is permitted for radiomedical advice.

Authority

The master of the vessel is in charge of all radio communication equipment regardless of who is operating it. Only the master can authorize transmission of the Two Tone Alarm or transmission of any DISTRESS, Urgency or Safety (Safety Of Life At Sea or SOLAS) procedures.

Radiotelephony Frequencies For SOLAS Procedures

2182 kHz	
4125 kHz	
6215 kHz	
8291 kHz	Reserved EXCLUSIVELY for DISTRESS and SAFETY
12290 kHz	
16420 kHz	
and VHF Channel 16 (FM)	

The frequencies listed above are also calling frequencies used for initial contact between ship-shore and ship-ship, which means these frequencies are watchkeeping frequencies.

In addition, the frequencies **156.8 MHz (Ch.16), 2182 kHz, 4125 kHz, 3023 kHz, 4125 kHz, 5680 kHz, 123.1 MHz,** and **156.3 MHz (Ch.6)** may be used for ship-aircraft communications.

Identification

All transmissions, whether bridge-bridge, bridge-shore, bridge-aircraft or on-scene communications (transmissions to be received on board the same vessel eg. bridge-bow) should be identified by use of the ship's name, radio callsign, or Maritime Mobile Service Identity (MMSI).

Bridge to Bridge Communications are inter-ship safety communications conducted by VHF radiotelephony from the position from which the vessel is normally navigated.

Modes of Transmission

Single Sideband Suppressed Carrier (SSB), known as J3E, is used for all radiotelephony communications on marine frequencies in the MF and HF Bands. The only exception to this is when Single Sideband with Full carrier, known as H3E, is used on 2182 kHz when in Distress. Use of H3E permits other stations to locate bearings by means of radio direction finding equipment.

Silence Periods are to increase the safety of life at sea. The Radio-telephony silence periods are from the hour to 3 minutes past and from 30 to 33 minutes past. They must be observed on the above Distress and Calling frequencies, including the DSC MF Distress frequency of 2187.5 kHz. Only Distress transmissions are permitted during these periods.

Time Signals (Now given in Coordinated Universal Time – UTC)
Time signals should be obtained daily to check accuracy of bridge chro-nometers, and to ensure correct observation of silence periods. Time signals may be obtained from stations included in the "List of Radiodetermination & Special Services" published by the International Telecommunications Union (ITU).

WWV/Fort Collins, Colorado transmits time signals on 5 MHz, 20 MHz and 15 MHz with a spoken announcement each minute. VNG/Llandilo, New South Wales transmits time signals on 2.5 MHz, 5 MHz, 8.638 MHz, 12.984 MHz and 16 MHz with a spo-ken announcement each minute.

RECEPTION OF MARITIME SAFETY INFORMATION (MSI)

Reception by NAVTEX

By consulting the appropriate documents watchkeepers should ascertain the designated NAVTEX areas they are transiting and select coverage accordingly. NAVTEX receivers have a LOGGING MODE which when enabled causes the first line (or PREAMBLE) of each received message to be printed out – regardless of whether the message emanates from a coast station which has been selected or not. The Logging Mode therefore is a good indicator of reception from various stations and may be useful to the watchkeeper in the selection of stations.

Message types are designated by codes (see page 71) and may be selected by use of keypads.

Reception by INMARSAT EGC

Ships selectively receive relevant SafetyNET (MSI) messages by logging in to the required ocean region and keeping their SES updated with the ship's position. The EGC MSI messages have Distress, Urgent, Safety or Routine priorities and if the position is not updated within 12 hours all messages above Routine priority, for the entire ocean region, will be received.

EGC for Coastal Warnings

International NAVTEX transmissions of coastal MSI are on 518 kHz. However as Australia's vast coastline of 37,000 km makes this method of broadcast impractical the NAVTEX system of MSI has been adapted for INMARSAT EGC use. The Australian coastline has been sectioned into areas A, B, C, D, E, F, G, H. To receive broadcasts for a particular area that area must be selected on the EGC receiver.

Coastal MSI covers areas which must be individually selected. Some warnings, however, are broadcast over adjoining areas when considered necessary.

Some SES equipment requires input of 'tracking' data. ie position information in latitude and longitude, for SafetyNet broadcasts.

EGC Addresses

EGC messages may be addressed to:-

a) Individual unique Identification (ID);
b) Group ID;

c) A pre-assigned geographical area (eg. for NAVAREA warnings); and

d) Absolute geographic areas may be circular or rectangular in shape. A circular area will be determined around a specific location (for example Latitude/Longitude). A rectangular area will be bounded by Northings and Eastings from the Southwest corner.

Once a message has been correctly received it will be automatically suppressed on subsequent broadcasts. Certain messages, however, will ALWAYS be received and printed, eg. initial reception of Distress Alerts and Navigational warnings.

Reception by HF NBDP

Navigational and weather information (MSI) is broadcast by some coast stations on the following NBDP frequencies:-
4210 kHz 6314 kHz 8416.5 kHz 12579 kHz 16806.5 kHz 22376 kHz 26100.5 kHz

HF Radiotelex is accessed through coastal radio stations using ARQ and FEC modes.

See ITU List of Radiodetermination & Special Services.

The Navigational Warning Signal of the Old Distress and Safety System
This consists of a single 2200 Hz tone of 250 milliseconds duration. This signal is transmitted by coast stations continuously for 15 seconds prior to vital navigational warnings on Radiotelephony.

The Navigational Warnings Transmitted by Radiotelephony are always preceded by an announcement on one or more of the Distress and Calling frequencies. The announcement consists of the Safety Signal *SECURITE* (pronounced *SAY-CURE-IT-AY*) spoken 3 times and which indicates that the calling station is about to transmit a message containing important navigational or important meteorological information. All stations hearing the Safety signal shall listen to the safety message until they are satisfied that the message is of no concern to them. They shall not make any transmission likely to interfere with the message. The call is addressed to all stations and includes specific working frequencies on which the MSI is to be broadcast. eg:-

SAYCURITAY SAYCURITAY SAYCURITAY

ALL SHIPS ALL SHIPS ALL SHIPS

THIS IS

STATION NAME STATION NAME STATION NAME

GALE WARNING FOLLOWS ON (working frequency)

All vessels hearing announcement now transfer to nominated working frequency

PROTECTION OF DISTRESS FREQUENCIES

Guard Bands
Any emission capable of causing harmful interference to Distress, Alarm, Urgency, or Safety communications on the frequencies
490 kHz, 518 kHz, 2174.5 kHz, 2182 kHz, 2187.5 kHz, 3023 kHz, 4125 kHz, 4177.5 kHz, 4207.5 kHz, 420-9.5 kHz, 4210 kHz, 5680 kHz ,6215 kHz, 6312 kHz, 6314 kHz, 8291 kHz, 8376.5 kHz, 8414.5 kHz, 8416.5 kHz, 12290 kHz, 12520 kHz, 12577 kHz, 12579 kHz, 16420 kHz, 16695 kHz, 16804.5 kHz, 16806.5 kHz, 19680.5 kHz, 22376 kHz, 26100.5 kHz, 121.5 MHz, 123.1 MHz, 156.3 MHz, 156.525 MHz, 156.650 MHz, 156.8 MHz, 406-406.1 MHz Band, 1530-1544 MHz Band, 1544-1545 MHz Band, 1626-1645.5 MHz Band, 1645.5-1646.5 MHz Band and 9200-9500 MHz Band, *which are identified for Distress, Alarm, Urgency or Safety is prohibited.*

Tests on Distress Frequencies shall be kept to a minimum, shall be coordinated with a competent authority and, as necessary and wherever practicable, be carried out on artificial antennas or with reduced power. Testing on the Distress and Safety calling frequencies should be avoided but where this is unavoidable it should be indicated that these are test transmissions.

Transmissions During Distress Traffic
The Distress call shall have absolute priority over all other transmissions. All stations which hear it shall immediately cease any transmission capable of interfering with the Distress traffic and shall

continue to listen on the frequency used for the emission of the Distress call. This call shall not be addressed to a particular station and acknowledgment of receipt shall not be given before the Distress message which follows it is sent.

Avoiding Harmful Interference

All stations shall radiate only as much power as is necessary to ensure a satisfactory service and special consideration shall be given to avoiding interference on distress and safety frequencies. Before transmitting for other than Distress purposes on any of the above frequencies, a station shall, where practicable, listen on the frequency concerned to make sure that no Distress transmission is being sent.

Prevention of Unauthorized Transmissions

All stations are forbidden to carry out:-

a) **unnecessary transmissions;**
b) **the transmission of superfluous signals and correspondence;**
c) **the transmission of false or misleading signals;**
d) **the transmission of signals without identification.**

Search And Rescue Operation (SAR)

THE ROLE OF RESCUE CO-ORDINATION CENTRES (RCCS)

National Mission Control Centres (MCCs) delegate SAR coordinating roles to their subordinate RCCs. A reliable communications link is necessary between MCC/RCC/Mobile Stations as it is entirely possible that Distress Alerts may be received and handled by RCCs many thousands of miles from the distress vessel.

MERCHANT SHIP SEARCH AND RESCUE (MERSSAR) MANUAL

This is published by the IMO to provide information and guidance on SAR operational techniques. Seafarers should make themselves familiar with the contents of this Manual.

MARITIME RESCUE ORGANIZATIONS operate in many countries. Some organisations are entirely funded by the government of the country concerned, and many more are voluntary organisations. The Australian Maritime Safety Authority (AMSA) is one such government organisation which provides a marine SAR service in Australian waters.

SHIP REPORTING SYSTEMS

Several countries operate ship reporting systems. Possibly the most well known of these is the United States Coast Guard Automated Mutual-Assistance VEssel Rescue System, otherwise known as the AMVER organisation. Ship reporting systems operated by other countries include JASREP in Japanese waters, SHIPPOS operated by Denmark in Danish Baltic waters and Brazil with SISCONTRAM.

The **Australian Ship Reporting (AUSREP) System** is a system which is compulsory for all vessels entering the region covered. Full details of **AUSREP** appear later in this publication. **(Appendix 5).**

Part 4

Miscellaneous Skills and Operational Procedures for General Communications

Ability to Use English Language, Both Written And Spoken, for the Satisfactory Exchange of Communications Relevant To the Safety Of Life at Sea

USE OF THE INTERNATIONAL CODE OF SIGNALS AND THE IMO STANDARD MARINE NAVIGATIONAL VOCABULARY/ SEASPEAK REFERENCE MANUAL.

Knowing how to operate GMDSS equipment is of little use if efficient radio communications cannot be established. The most widely used language for ship-shore-ship communication is English and proficiency both in written and spoken English is a basic necessity.

The IMO Standard Marine Navigational Vocabulary contains useful information for the Navigator on all aspects of Radiotelephony communication, including the following 'Message Markers':-

QUESTION	Indicates the following message is interrogatory
ANSWER	Indicates the following message is the reply to a question
REQUEST	Indicates the following message asks for action from others with respect to the ship
INFORMATION	Indicates the following message is restricted to observed facts
INTENTION	Indicates the following message informs others about immediate navigational actions intended to be taken
WARNING	Indicates the following message informs other traffic participants about dangers
ADVICE	Indicates the following message implies the intention of the sender to influence the recipient(s) by a recommendation
INSTRUCTION	Indicates the following message implies the intention of the sender to influence the recipient(s) by a regulation

Watchkeepers are required to understand and use the Vocabulary under the Certification Standards as set by the International Convention on Standards of Training, Certification and Watchkeeping for Seafarers, 1978.

The Seaspeak Reference Manual specialises in communication on VHF and comes with a pre-recorded audio-cassette tape which demonstrates relevant sections of the text. This useful publication, together with the IMO Standard Marine Navigational Vocabulary, is an attempt to standardize international maritime English. Seafarers should make themselves familiar with the contents of these manuals which are a lexicon of expressions intended to cover almost any situation which the Navigator may encounter whilst on the bridge.

RECOGNISED STANDARD ABBREVIATIONS AND COMMONLY USED SERVICE CODES

It is necessary for the operator to know the recognised standard abbreviations and codes, some of which may be found in the Manual for Use by the Maritime Mobile and Maritime Satellite Services, eg the SINPO (Signal Interference Noise Propagation Disturbance Overall rating). The SINPFEMO code provides for a more detailed radio reception report as it gives additional readings for Frequency and Modulation (quality and depth). Some commonly used codes are listed below:-

AA All After	Used after RPT to request or to give repetition of a message or part of a message eg. 'Repeat All After FUEL.'
AB All Before	'Repeat All Before ETA
BN All Between	Used to request (RQ) or to give a repetition eg. 'Request RPT BN ETA and pilot"
RQ Request	when requesting eg. RPT
BQ Reply to a Request	
Correction	Cancel my last word, here is the correct word – used to correct oneself when an error has been made in transmitting a message
ETA	Estimated Time of Arrival
ETD	Estimated Time of Departure
Kts	Knots
MSG	Message from/to a ships Master on ship's business
WX	Weather (message)
NX	Navigation warning/s
TR	Prefix indicating ship's name/position/ next port of call follows
DE This Is	Used when communications are difficult

USE OF INTERNATIONAL PHONETIC ALPHABET

Use of this Alphabet is standard world wide
(letters to be emphasised in italics)

Letter	Code word	Spoken as
A	ALFA	*AL* FAH
B	BRAVO	*BRAH* VOH
C	CHARLIE	*CHAR* LEE
D	DELTA	*DELL* TAH
E	ECHO	*ECK* OH
F	FOXTROT	*FOKS* TROT
G	GOLF	GOLF
H	HOTEL	*HOH* TELL
I	INDIA	*IN* DEE AH
J	JULIET	JEW LEE ETT
K	KILO	*KEY* LOH
L	LIMA	*LEE* MAH
M	MIKE	MIKE
N	NOVEMBER	NO *VEM* BER
O	OSCAR	OSS CAH
P	PAPA	PAH *PAH*
Q	QUEBEC	KEH *BECK*
R	ROMEO	*ROW* ME OH
S	SIERRA	SEE *AIR* RAH
T	TANGO	*TANG* GO
U	UNIFORM	*YOU* NEE FORM
V	VICTOR	*VIK* TAH
W	WHISKEY	*WISS* KEY
X	X-RAY	*ECKS* RAY
Y	YANKEE	*YANG* KEY
Z	ZULU	*ZOO* LOO
0	NADAZERO	NAH DAH ZAY ROH
1	UNAONE	OO NAH WUN
2	BISSOTWO	BEES SOH TOO
3	TERRATHREE	TAY RAH TREE
4	KARTEFOUR	KAR TAY FOWER
5	PANTAFIVE	PAN TAH FIVE
6	SOXISIX	SOK SEE SIX
7	SETTESEVEN	SAY TAY SEVEN
8	OKTOEIGHT	OK TOH AIT
9	NOVENINE	NO VAY NINER
.	DECIMAL	DAY SEE MAL
FULL STOP	STOP	STOP

Obligatory Procedures And Practices

EFFECTIVE USE OF OBLIGATORY DOCUMENTS AND PUBLICATIONS

It is imperative that candidates for the GMDSS GOC are aware of the documents on board and their uses.

RADIO RECORD KEEPING

A Radio Log is carried on board and must be kept up-to-date.

The Radio Log shall contain:-

a) a summary of communications relating to Distress, Urgency, and Safety traffic;

b) a reference to important service incidents (eg. MSI);

c) if the ship's rules permit, the ship's position at least daily; and

d) the time at which the above occurred.

KNOWLEDGE OF THE REGULATIONS AND AGREEMENTS GOVERNING THE MARITIME MOBILE SERVICE AND THE MARITIME MOBILE SATELLITE SERVICE.

Knowledge of the above-mentioned Regulations and Agreements is necessary for correct and efficient operation of GMDSS equipment and is contained in various publications, eg. The Manual for Use by the Maritime Mobile and Maritime Mobile-Satellite Services, the Admiralty List of Radio Signals Volumes 1-6, INMARSAT User's Manual, the NAVTEX Manual, etc.

Practical And Theoretical Knowledge Of General Communication Procedures

SELECTION OF APPROPRIATE COMMUNICATION METHODS IN DIFFERENT SITUATIONS

Operators should know which radiocommunications system will be the most suitable for any given ship's position. Factors to be considered when deciding which equipment to use may include time of day or night, radio wave propagation and the Ionosphere, facilities and services available at coast stations, price and ship's position.

TRAFFIC LISTS

Broadcast by coast radio stations on Radiotelephony (and, from some stations Radiotelex) frequencies, are lists consisting of ship's names, radio callsigns or MMSIs for which the coast station has traffic ie. radiotelephone calls, radiotelexes, radiotelegrams, etc.

RADIOTELEPHONE CALL

Method of Calling a Coast Station by Radiotelephony
Before transmitting, stations should take precautions to ensure emissions will not cause interference with transmission already in progress. The ship should call the coast station on a frequency on which it is keeping watch. The call shall consist of:-

NAME, RADIO CALLSIGN OR MMSI (ID) OF THE STATION CALLED (not more than 3 times)

THIS IS (or *DELTA ECHO* in case of language difficulties)

NAME, RADIO CALLSIG, OR MMSI (ID) OF THE CALLING STATION (not more than 3 times)

Ordering for a Manually Switched Link Call (known as Call Request)
Ship station then requests a channel for a manually switched telephone call. Alternatively the ship may initially call on a working channel assigned for radiotelephony on which the coast station is keeping watch. Frequencies may be ascertained from the ITU List of Coast Stations.

Controlling operator (at the coast station) then sets up the call and advises chargeable minutes on completion.

Ending the Call is indicated by use of the word **OUT** (or VA spoken as VICTOR ALFA in case of language difficulties).

Special Facilities of Calls

Calls may be cancelled by the calling station or refused by the called station.

Private calls are those other than Distress, Government (from government officials) or Service calls.

Station calls are those booked to a specific number and where the chargeable duration commences as soon as a reply is obtained at the called station, regardless of the person who answers the call.

Callers may request **information** about the telephone service.

By agreement, among the Administrations concerned, the following additional facilities may be granted:-

> **Personal calls** (when the call is to a specific person);
> **Data calls** (for the purpose of exchanging data between stations);
> **Collect calls** (when the caller requests the call to be paid for by the called party);
> **Credit-card calls** (when the caller requests the call be charged to a credit card number); and
> **Conference calls** (when a call is established between two or more stations).

There are, however, different types of Conference call:-

1) **Bidirectional** calls, in which each participant can listen and speak at will;
2) **Unidirectional** calls, in which only the leader can speak, and other participants are only able to listen; and
3) **Calls** which are a combination of the above.

Method of Calling a Coast Station by DSC

The call shall contain information indicating the station or stations to which the call is directed and the ID of the calling station. The call should also contain information indicating the type of communication to be set up and may include other information such as a working frequency or channel. This information shall always be indicated by coast stations which shall have priority for that purpose. The call shall be transmitted once on a single appropriate calling channel or frequency. Only in exceptional circumstances may a call be transmitted simultaneously on more than one frequency.

If the station called does not acknowledge the call it may be transmitted again on the same or another calling frequency after a period of at least 5 minutes (5 seconds in automated VHF or UHF systems). It should then normally not be repeated until after a further interval of 15 minutes.

When initiating a call to a coast station, a ship station should preferably use the coast station's nationally assigned calling channels, for which purpose it shall send a single calling sequence on the selected channel.

A typical DSC calling and acknowledgment sequence contains the following signals:-

Signal	Method of composition
Format specifier	selected
Address	entered
Category	selected
Self-ID	pre-programmed
Telecommand information	selected
Frequency/channel (if appropriate)	entered
Telephone number (semi-automatic/automatic	
Ship-shore connections only	entered
End of sequence signal	selected

Selecting an Automatic Radiotelephone Call

See format above

RADIOTELEGRAM

The Parts of a Radiotelegram

The Preamble consists of :-

    Name of Office of Origin;

    Number of chargeable/actual words;

    Date & Time of handing in; and

    Service instructions (if any).

Service Instructions and indications are instructions added to a telegram by the Office of Origin or by another office:-

Ampliation (telegram sent a second time);

URGENT (Urgent telegram); and

SVH (telegrams referring to the safety of life in cases of exceptional urgency).

Address must contain all particulars necessary to effect delivery without enquiry or request for information.

Text

Each telegram must contain a text that contains at least one character. The text must be continuous (no blank lines etc) and must be written in characters used in the country of origin and which have an equivalent in the standard Latin alphabet and figures 1 to 0. The following punctuation signs etc may be used:- . , : ? ' + − / = ( ).

Signature is not compulsory. It may be written by the sender in any form.

Addresses

Full addresses must include the designation of addressee, name and number (if available) of street, name of office of destination and name of telegraph office of destination with postal code (in brackets). If street/number is unavailable, then occupation of addressee. Surnames shall be accepted as the sender writes them and other particulars of address written in the language of country of destination. The name of the office of destination should be written in accordance with the relevant columns of the List of Telegraph Offices.

A **Registered Address** is one in which the full address (including the office of destination) is replaced by a single simplified indication.

A **Telephonic Address** is prefixed with Tfx (where x is the telephone number), appears as first word of the address (with if necessary, name or access number of network), followed by name of addressee and office of destination (eg town).

A **Telex Address** is prefixed with TLXx (where x is the telex number) and appears as the first word of the address followed by name of addressee and office of destination.

Counting of Words

Words, groups of characters or single characters not exceeding 10 characters, shall be counted as one word. In excess of 10 characters shall be counted as 1 chargeable word for each 10 characters or part thereof. For example RADIONAVIR is one word but RADIONAVIRE whilst one actual word is chargeable as two, and as the chargeable number always appears first appears in the preamble as 2/1.

Transmission of a Telegram by Radiotelephony by Stations of the Maritime Mobile Service should include routine repetitions of, for example, difficult words etc. In the case of loss of communication with the receiving station and when communication is not envisaged to re-commence within 24 hours, the sender must be informed of the reason for non-transmission of the radiotelegram.

Transmission of a Telegram by Radiotelex Stations of the Maritime Mobile Service should normally transmit and receive radiotelegrams by means of radiotelex only.

TRAFFIC CHARGES

AAIC Code

Every mobile station has an Accounting Authority Identification Code (AAIC) which identifies the accounting authority responsible for the settlement of maritime accounts.

International Charging System

Charges for radio communications from ship to shore shall in principle and, subject to national law and practice, be collected from the maritime mobile station licensee by:-

a) the Administration that has issued the licence; or

b) a recognised private operating agency; or

c) any other entity or entities designated for this purpose by the Administration that has issued the licence.

Charges for specific services may be found in the ITU List of Coast Stations.

INMARSAT Communications Charging System

Charges for services may be found in the INMARSAT Users Handbook.

Land Line Charge (LLC) is the charge relating to transmission over the general network of telecommunication channels, national and international.

Coast Charge (CC) is the charge relating to the use of facilities provided by the land station in the maritime mobile services or by the earth station in the Maritime Mobile-Satellite Service. In the Maritime-Mobile Satellite Service this charge shall include all space segment costs. An Administration may also choose to present its total land station charge in its component parts.

Ship Charge (SC) is the charge collected on board by the mobile station relating to the use of facilities provided by the mobile station.

Currencies Used in International Charging will be either the monetary unit of the International Monetary Fund known as the Special Drawing Right **(SDR)** or the Gold Franc **(GF)**. **(1 GF = 3.061 SDR)**.

PRACTICAL TRAFFIC ROUTINES

Candidates for the GMDSS GOC to be given practical experience in the exchange of radio communications traffic eg. in Radiotelephony, Radiotelex, INMARSAT-A, INMARSAT-B, or INMARSAT-C.

World Geography, Especially the Principal Shipping Routes and Related Communications Routes.
Students are to be made aware of the major trade routes for shipping and the facilities available world wide for different means of communications. In this way the best means of communication, for any given geographical ship position, may be readily ascertained.

APPENDIX 1

EXAMINATION SYLLABUS
(Recommendation T/R 31-03E (Bonn 1993))

EXAMINATION SYLLABUS

(Recommendation T/R 31-03E (Bonn 1993)) The examination should consist of theoretical and and practical tests and shall include at least:-

A. KNOWLEDGE OF THE BASIC FEATURES OF THE MARITIME MOBILE SERVICE AND THE MARITIME MOBILE SATELLITE SERVICE

A1 The general principles and basic features of the Maritime Mobile Service

A2 The general principles and basic features of the Maritime Mobile Satellite Service

B. DETAILED PRACTICAL KNOWLEDGE AND ABILITY TO USE THE BASIC EQUIPMENT OF A SHIP STATION

B1 Use in practice the basic equipment of a ship station

B2 Digital Selective Calling (DSC)

B3 General principles of Narrow Band Direct Printing (NBDP) and Telex Over Radio (TOR) systems. Ability to use maritime NBDP and TOR equipment in practice

B4 Usage of INMARSAT systems, INMARSAT equipment or simulator in practice

B5 Fault locating

C. OPERATIONAL PROCEDURES AND DETAILED PRACTICAL OPERATION OF GMDSS SYSTEM AND SUBSYSTEMS

C1 Global Maritime Distress and Safety System (GMDSS)

C2 INMARSAT

C3 NAVTEX

C4 Emergency Position Indicating Radio Beacons (EPIRBs)

C5 Search and Rescue Radar Transponder (SART)

C6 Distress, urgency, and safety communication procedures in the GMDSS

D. MISCELLANEOUS SKILLS AND OPERATIONAL PROCEDURES FOR GENERAL COMMUNICATIONS

D1 Ability to use English language, both written and spoken, for the satisfactory exchange of communications relevant to the safety of life at sea

D2 Obligatory procedures and practices

D3 Practical and theoretical knowledge of general communication procedures

APPENDIX 2

RADIOTELEPHONY IN PRACTICE

RADIOTELEPHONY IN PRACTICE

It is of little using knowing how to operate GMDSS equipment if upon picking up a microphone the operator does not know what to say.

Maritime Communications Operators are able to call stations on the correct frequencies, change to working frequencies or channels, pass information quickly and efficiently with minimum need for repetition and without causing interference to other stations. The importance of clear, efficient communication cannot be overstated.

Decide what you are going to say before you pick up the microphone.When passing information it is often useful to write it down first.

Speak clearly and slowly and do not hold the microphone too close as this will cause distortion.

Words and Phrases In Common Usage

Word/Phrase	Meaning
ROMEO	Received
STANDBY	Wait a short time
SAY AGAIN	Repeat
WORD AFTER WORD BEFORE ALL AFTER ALL BEFORE	All of these phrases refer to a request for information or to a repeat of information eg. "Say again Word After FUEL" or "I repeat All Before PILOT"
CORRECTION	Used to correct oneself eg "ETA – CORRECTION – ETD"
OVER	Go ahead. Transmit now (must be used on Simplex)
OUT	End of transmission
SIMPLEX	One-way speech (single frequency used for transmit/receive)
DUPLEX	Normal conversation (separate frequencies use for transmit/receive)
I SPELL	I am about to give the phonetic spelling of the word I have just read
DECIMAL	Decimal point
STOP	Full stop

"HOW DO YOU READ ME?"	"How well do you understand me?"
Readability	**Signal Strength**
I READ YOU 1/Bad	1/Barely perceptible
2/Poor	2/Weak
3/Fair	3/Fairly good
4/Good	4/Good
5/Excellent	5/Very good

EXAMPLES OF MESSAGE TRANSMISSION

When passing information it is often useful to give the phonetic spelling of difficult words and phrases in order to facilitate accurate reception.

We may do this by first saying the difficult word then saying the words **I REPEAT** or **I SPELL**, then spelling out the word phonetically.

'Word' **means a recognised word, a letter or figure by itself or a group of mixed letters and figures. In short, any individual part of a message whether in address, text or signature.**

Figures, whether by themselves or in groups, letters by themselves or in code groups, or groups of mixed figures and letters should ALWAYS be spelt out.

1) *ETA PILOT GROUNDS 151500Z I SPELL 151500 ZULU*

2) *REQUEST PART NUMBER BA/1940.SP I SPELL BRAVO ALFA OBLIQUE STROKE ONE NINER FOUR ZERO DECIMAL SIERRA PAPA*

3) *SAY AGAIN ALL AFTER SIX I SPELL SIERRA INDIA XRAY*

4) *REQUIRE 26 I SPELL BEESOHTOO SOKSEESIX TONNES*

SUMMING UP, SAY THE DIFFICULT WORD OR GROUP, THEN THE WORDS *I SPELL*, THEN GIVE THE PHONETIC SPELLING SLOWLY AND CLEARLY.

ESTABLISHING COMMUNICATIONS
To minimize interference always listen first before calling and remember that every call must be identified.

Whether to a coast radio station or to another ship the initial call will always be made on one of the International Distress Safety and Calling frequencies. Once contact has been established both stations then transfer to a working frequency.

Because coast station operators are usually listening to many different frequencies, it is good practice to state your frequency and the reason for your call. The coast station operator then knows the best frequency to transfer you to and can make your traffic arrangements more speedily.

MF RADIOTELEPHONY
Medium Frequency (MF) for Radiotelephony purposes is the band of frequencies between 2 MHz and 3 MHz.

HF RADIOTELEPHONY
High Frequency (HF) Radiotelephony is the Band of frequencies from 3 MHz to 30 MHz.
Initial contact is made on one of the following DISTRESS and Safety frequencies

kHz

2182

4125

6215

8291   this frequency reserved EXCLUSIVELY for Distress and Safety

12290

16420

Channel 16 VHF

The initial call consists of the *ID* (NAME, RADIO CALLSIGN OR MMSI) of the called station not more than 3 times followed by

this is

followed by

ID of the calling station not more than 3 times.

1) HF 4125 kHz

DARWINRADIO DARWINRADIO
THIS IS
TAIPAN TAIPAN ON 4125 kHz REQUEST RADFONE CALL ON
CHANNEL 419 OVER

TAIPAN TAIPAN
THIS IS
DARWINRADIO DARWINRADIO

ROMEO CHANNEL 419

ROMEO

both stations now change frequency to channel 419 for radiotelephone call

DARWINRADIO DARWINRADIO
THIS IS TAIPAN TAIPAN

HOW DO YOU READ ME?
OVER

TAIPAN TAIPAN
THIS IS
DARWINRADIO DARWINRADIO
READING YOU FOUR TO FIVE.
WHAT IS YOUR ALFA ALFA
INDIA CHARLIE (accounting code)
AND WHAT TELEPHONE
NUMBER YOU REQUIRE?
OVER

However, on VHF and in good conditions, the call may consist of the called station ID *once* only followed by

this is

ID of calling station (twice)

2) VHF Channel 16

DARWINRADIO
THIS IS
TAIPAN TAIPAN ON VHF CHANNEL 16 POSITION REPORT
OVER

> TAIPAN
> THIS IS
> DARWINRADIO DARWINRADIO
> CHANGE TO CHANNEL 67
> OVER

ROMEO

both stations now change frequency to VHF channel 67

DARWINRADIO
THIS IS
TAIPAN TAIPAN ON CHANNEL 67 HOW DO YOU READ ME?
OVER

> TAIPAN
> THIS IS
> DARWINRADIO READING YOU
> FIVE FIVE GO AHEAD WITH
> YOUR PAPA ROMEO
> OVER

3) VHF Channel 16

SINGAPORERADIO
THIS IS
SUNRISER SUNRISER VHF CHANNEL 16 WITH NAVIGATION
* WARNING*
OVER

> SUNRISER
> THIS IS
> SINGAPORERADIO
> SINGAPORERADIO CHANGE TO
> CHANNEL 4 I REPEAT
> CHANNEL 4
> OVER

ROMEO

both stations now change to VHF channel 4

SINGAPORERADIO
THIS IS
SUNRISER SUNRISER HOW DO YOU READ ME?
OVER

> SUNRISER
> THIS IS
> SINGAPORERADIO
> SINGAPORERADIO READING
> YOU FIVE FIVE GO AHEAD
> OVER

SINGAPORERADIO
THIS IS
SUNRISER
NAVIGATION WARNING
WE ARE 22000
I REPEAT *– TWO TWO ZERO ZERO ZERO – TONNE TANKER*
  DRIFTING IN POSITION LATITUDE ZERO ZERO ONE FIFE
  NORTH LONGITUDE ONE ZERO FOUR FIFE EAST UNTIL
  APPROX 161500Z
I REPEAT *– ONE SIX ONE FIFE ZERO ZERO ZULU – DUE MAIN*
  ENGINE REPAIRS
OVER

> SUNRISER
> THIS IS
> SINGAPORERADIO REPEAT
> WORD AFTER ONE SIX ONE
> FIFE ZERO ZERO ZULU
> OVER

SINGAPORERADIO
THIS IS
SUNRISER
WORD AFTER ONE SIX ONE FIFE ZERO ZERO ZULU DUE
I SPELL – DELTA UNIFORM ECHO
I REPEAT   DELTA UNIFORM ECHO
OVER

SUNRISER
THIS IS
SINGAPORERADIO
RECEIVED YOUR NAVIGATION
WARNING.
REQUEST YOU ADVISE WHEN
REPAIRS COMPLETED
OVER

SINGAPORERADIO
THIS IS
SUNRISER WILL ADVISE.
OUT

ROMEO OUT

In a case such as the one above, the vessel SUNRISER should already have broadcast an announcement addressed to All Stations on Channel 16, with the actual Safety message being transmitted on Channel 13.

4.   VHF Channel 13 after DSC Alert on Channel 70

DSC SAFETY CALL ON VHF CHANNEL 70 ADDRESSED TO ALL SHIPS AND SPECIFYING WORKING CHANNEL 13

On VHF Channel 13

SAYCURITAY SAYCURITAY SAYCURITAY
ALL SHIPS ALL SHIPS ALL SHIPS
THIS IS
OLIVEBANK OLIVEBANK OLIVEBANK
232000037
NAVIGATION WARNING
AT 161500ZULU SIGHTED PARTLY SUBMERGED CONTAINER
    BEARING 220 DEGREES 7 MILES OFF USHANT
I SAY AGAIN
AT 161500ZULU SIGHTED PARTLY SUBMERGED CONTAINER
    BEARING 220 DEGREES 7 MILES OFF USHANT.
WIND SOUTHWESTERLY 27 KNOTS WITH HEAVY WESTERLY
    SWELL
OUT.

5. DSC URGENCY CALL ON HF 16804.5 kHz ADDRESSED TO PERTHRADIO COAST RADIO STATION AND SPECIFYING WORKING CHANNEL 1602

On HF Channel 1602

PANNE PANNE PANNE PANNE PANNE PANNE
PERTHRADIO PERTHRADIO PERTHRADIO
THIS IS
BIENVENIDOS BIENVENIDOS   BIENVENIDOS
755000079
RADIOMEDICAL
OVER

6. DSC DISTRESS ON 2182 KHZ AFTER DSC DISTRESS ALERT ON 2187.5 kHz

On 2182 kHz

RADIOTELEPHONE TWO TONE ALARM SIGNAL SENT FOR 45 SECONDS

MAYDAY MAYDAY MAYDAY   Distress Call
THIS IS
HARLEQUIN FOXTROT ALFA ALFA EX-RAY
HARLEQUIN FOXTROT ALFA ALFA EX-RAY
HARLEQUIN FOXTROT ALFA ALFA EX-RAY

MAYDAY	**Distress Signal**
HARLEQUIN FOXTROT ALFA ALFA EX-RAY	**Name/Callsign**
POSITION 12 MILES NORTHEAST OF CAPE CORDILLERA	**Position**
I AM ON FIRE IN THE ENGINE ROOM	**Nature of Distress**
12 CREW ON BOARD	**Other information**
2 LIFERAFTS I SAY AGAIN MY POSITION	**which may**
12 MILES NORTHEAST OF CAPE CORDILLERA	**facilitate rescue**
OVER	

MAYDAY
HARLEQUIN HARLEQUIN HARLEQUIN
FOXTROT ALFA ALFA EX-RAY-
THIS IS RELIANT RELIANT RELIANT
GOLF GOLF BRAVO ZULU
ROMEO ROMEO ROMEO MAYDAY
OVER

MAYDAY
RELIANT RELIANT RELIANT
THIS IS
HARLEQUIN HARLEQUIN HARLEQUIN
ROMEO
I AM ABANDONING SHIP I REQUEST
IMMEDIATE ASSISTANCE I HAVE
12 CREW 2 LIFERAFTS 2 SARTs
OVER

MAYDAY
HARLEQUIN HARLEQUIN HARLEQUIN
THIS IS
RELIANT RELIANT RELIANT
GOLF GOLF BRAVO ZULU
I HAVE LOCATED YOU ON MY RADAR
I AM COMING TO YOUR
ASSISTANCE MY ETA IN 45 MINUTES
I SAY AGAIN MY ETA IN 45 MINUTES
OVER

**RELIANT REPEATS MESSAGE TWICE AND RECEIVING NO
ACKNOWLEDGMENT ASSUMES CONTROL OF DISTRESS
COMMUNICATIONS**

RELIANT RELIANT RELIANT
THIS IS
SAPPHIRE SAPPHIRE SAPPHIRE
OVER

*SEELONCE MAYDAY (**from RELIANT**)*

RELIANT NOW ACCESSES DSC CONTROLLER RECEIVED DISTRESS MESSAGES AND RETRANSMITS THE RECEIVED ALERT AS A DSC DISTRESS RELAY CALL ON 2187.5 kHz, ADDRESSED TO ALL STATIONS

On 2182 kHz

RADIOTELEPHONE TWO TONE ALARM SIGNAL SENT FOR 45 SECONDS

MAYDAY RELAY MAYDAY RELAY MAYDAY RELAY
THIS IS
RELIANT RELIANT RELIANT
GOLF GOLF BRAVO ZULU

MAYDAY
FOLLOWING RECEIVED FROM
HARLEQUIN FOXTROT ALFA ALFA EX-RAY AT 230155Z
MAYDAY
HARLEQUIN FOXTROT ALF ALFA EX-RAY
POSITION 12 MILES NORTHEAST OF CAPE CORDILLERA
I AM ON FIRE IN THE ENGINE ROOM 12 CREW ON
BOARD 2 LIFERAFTS I SAY AGAIN MY POSITION
12 MILES NORTHEAST OF CAPE CORDILLERA
I AM ABANDONING SHIP I REQUEST IMMEDIATE
ASSISTANCE I HAVE 12 CREW 2 LIFERAFTS 2 SARTs
THIS IS RELIANT GOLF GOLF BRAVO ZULU
OVER

> MAYDAY
> RELIANT RELIANT RELIANT
> GOLF GOLF BRAVO ZULU
> THIS IS
> SAPPHIRE SAPPHIRE SAPPIRE
> VICTOR JULIET ALFA DELTA
> RECEIVED MAYDAY RELAY
> I AM 2000 TONNE CARGO
> VESSEL MY POSITION 6 I SAY
> AGAIN 6 MILES WEST OF
> HARLEQUIN
> INTENTION I AM PROCEEDING
> TO ASSIST MY ETA IS 1 HOUR
> OVER

MAYDAY
SAPPHIRE SAPPHIRE SAPPHIRE
VICTOR JULIET ALFA DELTA
THIS IS
RELIANT RELIANT RELIANT
GOLF GOLF BRAVO ZULU
ROMEO
I HAVE LOCATED YOU ON MY RADAR
I AM 58000 TONNE BULK CARRIER
INTENTION I WILL MAKE A LEE FOR
YOUR VESSEL AT DISTRESS SCENE
OVER

MAYDAY
RELIANT RELIANT RELIANT
GOLF GOLF BRAVO ZULU
THIS IS
SAPPHIRE SAPPHIRE SAPPHIRE
VICTOR JULIET ALFA DELTA
ROMEO

RELIANT NOW ACCESSES DSC CONTROLLER RECEIVED DISTRESS MESSAGES, AND RETRANSMITS THE RECEIVED ALERT AS A DSC DISTRESS RELAY CALL ON 2187.5 kHz, ADDRESSED TO ALL STATIONS

On 2182 kHz

RADIOTELEPHONE TWO TONE ALARM SIGNAL SENT FOR 45 SECONDS

MAYDAY RELAY MAYDAY RELAY MAYDAY RELAY
THIS IS
RELIANT RELIANT RELIANT
GOLF GOLF BRAVO ZULU

MAYDAY
FOLLOWING RECEIVED FROM
HARLEQUIN FOXTROT ALFA ALFA EX-RAY AT 230155Z
MAYDAY
HARLEQUIN FOXTROT ALFA ALFA EX-RAY
POSITION 12 MILES NORTHEAST OF CAPE CORDILLERA

I AM ON FIRE IN THE ENGINE ROOM 12 CREW ON
BOARD 2 LIFERAFTS I SAY AGAIN MY POSITION
12 MILES NORTHEAST OF CAPE CORDILLERA
I AM ABANDONING SHIP I REQUEST IMMEDIATE
ASSISTANCE I HAVE 12 CREW 2 LIFERAFTS 2 SARTs

THIS IS RELIANT GOLF GOLF BRAVO ZULU
VESSEL SAPPHIRE VICTOR JULIET ALFA DELTA PROCEEDING
TO ASSIST ETA DISTRESS SCENE 0230 ZULU
VESSEL RELIANT GOLF GOLF BRAVO ZULU ALSO PROCEEDING
TO ASSIST ETA DISTRESS SCENE 0245 ZULU
OVER

**(ACKNOWLEDGEMENTS RECEIVED FROM OTHER
VESSELS WHO ARE UNABLE TO ASSIST)**

MAYDAY
RELIANT RELIANT RELIANT
THIS IS
SAPPHIRE SAPPHIRE SAPPHIRE
I HAVE RESCUED 12 I SPELL –
ONE TWO – SURVIVORS
ALL SURVIVORS RESCUED
OVER

MAYDAY
SAPPHIRE SAPPHIRE SAPPHIRE
VICTOR JULIET ALFA DELTA
THIS IS
RELIANT RELIANT RELIANT
GOLF GOLF BRAVO ZULU
RECEIVED YOUR MESSAGE ALL
12 SURVIVORS RESCUED
OUT

MAYDAY
HELLO ALL STATIONS HELLO ALL STATIONS HELLO ALL STATIONS
THIS IS
RELIANT GOLF GOLF BRAVO ZULU
090420 ZULU **(time of handing in of the message)**
HARLEQUIN FOXTROT ALFA ALFA EX-RAY
(station which was in Distress)

SEELONCE FEENEE

APPENDIX 3

PUTTING IT ALL TOGETHER

PUTTING IT ALL TOGETHER

The aim of this section is to help the Watchkeeper make the right choices – every second counts.

Night time. Another uneventful watch on a long trip. The Navigator paces the bridge, while he thinks about home and looks forward to a shower and a good sleep.

Visibility has been poor all night, only about 2 miles due to a thick bank of fog, and the Navigator suddenly realises the fog has closed right in and he can only just make out the for'ard mast.

Going over to the radar he sees it has been left on the 4 mile range. Switching to 12 mile range he sees a target closing at 5 miles.

What now?

There is a full range of GMDSS equipment on board.

Does he send a DSC Alert? If so should it be sent on VHF Channel 70 or 2187.5 kHz or both?
Does he attempt to contact the target on VHF Channel 16?
Does he alter course?
Does he inform the Master?

But what does he do first?

Circumstances can change very rapidly at sea, or in harbour for that matter. Even if everything is done correctly and by the book, a short lapse of concentration, a negligent watchkeeper on board another vessel, perhaps even sheer bad luck, may create a dangerous situation.

A number of scenarios are presented in the following pages. Navigators are invited to consider them as exercises, the aim of which is to increase the Safety Of Life At Sea by promoting alertness and awareness.

Address each of the scenarios presented with the points below in mind.

● If you decide to send an Alert which equipment will you use?

● If you decide to make a broadcast or to contact another station, remember you are about to occupy a radio frequency and that every transmission must be identified.

● Listen first and make sure you will not cause interference to other stations.

● Think about what you are going to say on air, write it down if necessary.

● Make sure you know which station you are talking to.

1.

It is night and your vessel, the 30,000 GRT bulk carrier TRITON/ GVCX, is at anchor in company with 25 other vessels. Helicopter GAVBM is approaching to evacuate a sick crew member from your ship. How will you contact the helicopter pilot and how can you help him single out your vessel?
What information will the pilot require from you prior to landing?
How will you ready the deck for landing?
What safety precautions will you put in place?

2.

You are 2000 GRT cargo vessel MONTEVERDE/3ETCG and sight a derelict in position 36.37.22 South 57.03.23 West.

3.

You are 20,000 GRT tanker RASTANURA/VCCM approaching the fairway off Sharjah.
The tug OCEANUS is in the fairway towing a dredge across your course.
There has been no MSI from Gulf Port Control radio station.

4.

Your vessel the 7,000GRT gas tanker ARIADNE/SXYM is leaving port when you see a red flare approximately 270 degrees 2 miles distant.

5.

Your ship is the 11,000 GRT refrigerated cargo vessel HARDWICKE GRANGE/GHMU. You have cleared a bend in the channel and your radar shows an unknown vessel approaching on the port side about 8 miles ahead.
The vessel is expected to pass safely.
The vessel SILVER MOON is astern, has not cleared the bend and is also approaching to overtake on the port side.

6.

You are 1500 GRT coastal cargo vessel ZULEIFA/9XTZ and see that 15,000 GRT gas tanker ORIENT MAID is running in to shallow water.

7.

You are 9,000 GRT timber cargo vessel WAVE ISLAND/6ZHY with a seriously injured crew member.

8.

You are the 6,000 GRT cargo vessel GRYPHON/ZCVA and are proceeding in approach channel with sister ship UNICORN.

You are called by UNICORN/ZCVB whose steering gear does not respond and who has stopped engines to investigate.

You see that UNICORN is drifting into the wrong channel.

9.

You are 28.000 GRT bulk carrier LA ESTANCIA/GVQB and receive a DSC Alert on Channel 70.

Listening on Channel 16 you hear a Safety broadcast from ocean going tug STELLA /PFHD which has lost the difficult tow GLOGAS 3/C2EG drilling platform, in heavy seas, 2 miles ahead.

10.

Your vessel the 14,000GRT WAVE/6TZA is in collision with a yacht which is sinking.

11.

You are 60,000 GRT container vessel ATLANTIC CHIEF/SVVA in Southbound Channel off Red Cape.

Deck containers have shifted and angle of loll is 18 degrees.

Vessel is not answering helm.

12.

You are 20,000 GRT chemical tanker OCEAN CHEMIST/LDLB with poisonous cargo and have grounded on Clam Shoal.

You may float off with the tide if you jettison cargo.

If you do not jettison cargo there is a risk that the ship will heel over as the tide ebbs and cargo could be lost if tanks are ruptured.

13.

You are trawler ADELE/VM5775 in thick fog and see unknown vessel on radar running across your trawl.

14.

Your ship is the 15,000 GRT tanker OLYMPIC CITY/SWBB position 197 degrees 3 miles off Medusa Rock.

You are towing disabled cargo vessel 2,000 GRT MYKONOS/SXYT into approach channel when your propeller fouls on wire cable.

15.

You are 600 GRT fishing vessel AGHIOS NIKOLAOS/SW5182 and see that OLYMPIC CITY is standing into danger. There is a submerged wreck in position 42.35.17 North 21.37.22 East.

16.

You are 19,000 GRT products tanker JANJAN/5BTE with a full cargo. The 3rd Engineer reports on the internal telephone that there is fire in the engine room.

He requests all hands to the engine room to attempt extinguish the fire.

APPENDIX 4

SAFETY EDUCATION ARTICLE 64
Reproduced by kind permission of the
Australian Maritime Safety Authority
PO Box 1108 Belconnen
ACT 2616 Australia

CARE, MAINTENANCE AND INSTALLATION OF HF MARINE RADIO TRANSCEIVERS - TECHNICAL NOTES

INTRODUCTION

This SEA should not be interpreted by readers as a guide to installing a HF marine transceiver. It is important for readers to appreciate that maximum efficiency and effectiveness of a HF marine radio will only be achieved if it is professionally installed. Rather, this SEA is a guide so that operators are aware of the important components, aspects and issues within HF installations. This knowledge will help operators to maintain their marine radios to a high level of effectiveness.

IMPORTANT ASPECTS OF A HF INSTALLATION

The components of a HF marine transceiver which need to be examined are:

- Power supply
- Antenna (or aerial)
- Grounding
- Radio and electrical interference
- Non-electrical aspects

A typical system along with the block diagram is shown in fig below.

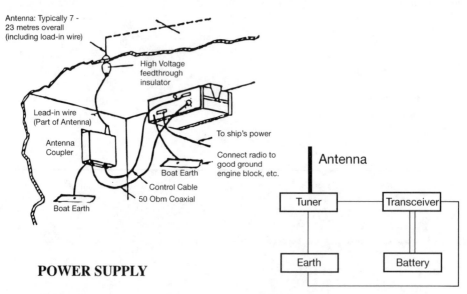

POWER SUPPLY

Battery

Ensure that the battery has sufficient capacity for your requirement. A 12 volt car battery rated at 40 to 60 amp hours is suitable for the lower power installations (round 100 - 125 watts). However in an emergency all unnecessary lighting and appliances should be switched off, to conserve power for transmitting. Larger installations should adhere to the Uniform Shipping Laws code, Section 9, paragraph 27.12

Charging Unit

The charging system needs to have sufficient capacity and be in good repair to keep the battery(s) fully charged at all times. It is wise to ensure your charging system can fully charge your batteries from a flat state in around 12 hours. For a 60 amp hour battery supply this means the charging system must, at the very least, be able to supply 7 to 8 amps, in addition to the craft's operating requirement, for charging the battery. For a small craft with a 60 amp hour battery, a standard car alternator system, kept in good condition, is more than satisfactory. But be warned, car systems were not designed for the harsh marine environment and may therefore quickly degrade into a less than satisfactory state.

Wiring

A correctly installed HF transceiver should be connected directly to the battery supply with as short a length of power cable as possible and not via complicated patch panels and other switch boxes as this will add unwanted resistance and downgrade the performance of your set.

All wiring needs to be in good condition, free of corrosion and clearly identifiable for ease of maintenance. The wiring between the battery and the transceiver should be of sufficient current carrying capacity. Wiring capacity is important for two reasons: firstly, if the wire is too thin it will not be able to supply sufficient power to the equipment to operate it correctly, and secondly, drawing too much current through low capacity wire is a fire hazard.

On all installations the battery cable should be fused at the battery end to protect the wiring. For lower power installations (around 100 - 125 watts) this should be with a 32 Amp fuse. In addition, the wire used to connect the transceiver to the battery on lower powered installations should have a cross section of at least 2.0 sq. mm if the run is reasonably short (i.e. about 3 metres) and greater than 3.0 sq. mm if the run is in excess of 5 metres. Specialists in the marine industry recommend cable of 4 to 6 sq. mm. This cable is more than adequate and will provide superior performance over the life of the system. Ideally, use specialised double insulated multi core marine cable which is purpose-designed for the marine environment.

(Note: As a rule the power supply cable voltage drop should never exceed 0.5 volts (Transmitter on) for all size installations.

ANTENNAS

Antenna

System Antennas should be installed as far away from other metal structures as possible. Additionally, the antenna feed (the connection between the tuner and the antenna base) must be kept as short as possible to avoid power loss and thereby maximising transmission efficiency. One method to ensure this is to mount the antenna tuner at the base of the antenna. However, if a long antenna feed from the transmitter is unavoidable, (as is likely in vessels with a manual tuner) the use of the correct type of wire will help minimise the losses and lessen the threat of RF (radio frequency) burns. The use of an automatic tuner enables it to be located close to the antenna.

The feed to the antenna must be waterproof to avoid corrosion of the current carrying element and the associated reduction of efficiency. Regular inspection and cleaning off of salt build-up on porcelain insulators is also necessary to ensure the continuing efficiency of the antenna system.

Antenna Polarity

Horizontal antennas are better for short to medium range skywave communications, while vertical antennas are better for long distance skywave and ground wave communications.

The vertical whip antenna produces vertically polarised radiation. The "back-stay" antenna, used in most yachts, is good because it provides a compromise between horizontal and vertical polarisation. Additionally its length makes it quite efficient in the low to medium frequency range (2 to 8 MegaHertz (MHz)).

Connecting the Feed to a Backstay Antenna

Vertical whip antennas are simply connected to the end of the antenna feed, generally by a screw thread. The fitting of a backstay antenna to its antenna feed is not so simple. It must not cause any interference or hazard to the crew, but must be accessible and watertight.

A popular technique used by many professional installers is shown in the figure opposite. By turning the feed back on itself, water will tend not to work up into the cable and corrode the wires. This technique will help ensure a long life for the installation.

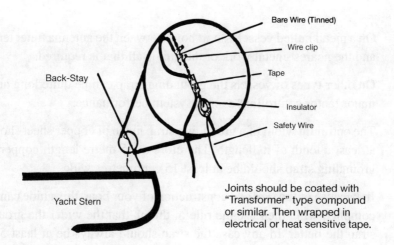

Bare Wire (Tinned)

Wire clip

Back-Stay

Tape

Insulator

Safety Wire

Yacht Stern

Joints should be coated with "Transformer" type compound or similar. Then wrapped in electrical or heat sensitive tape.

GROUND ASPECTS

Grounding

An effective ground for a HF transmitter is probably the most overlooked aspect of a HF installation but it is vitally important for the efficient performance of the system.

Grounding systems can be grouped into two categories:

Resistive grounds which rely on direct contact between a grounding plate, or the hull of a metal hulled vessel, and the sea and;

Capacity grounds where the grounding strap is attached to the keel bolt which is effectively insulated from the sea by the epoxy coating or paint of the keel.

There is little evidence to suggest that one grounding type is superior to the other in principle. It is the way the system is wired and maintained which will dictate its performance. However resistive grounds such as "dyna" plates should be carefully installed and always kept free of corrosion, algae, marine encrustations and especially paint as these all reduce the effectiveness of the ground and hence the HF installation. It is important to note that a good coat of paint will make the plate useless as a ground connection resulting in a catastrophic effect on the HF installation.

Grounding Straps.

The connection between the antenna tuner and the grounding point is called the grounding strap. The grounding strap should be made of copper sheet or copper pipe, not simply a wire.

On a metal hulled vessel a short bond between the antenna tuner terminal and the nearest metal work of the hull is all that is required.

On other types of vessels the grounding strap can be quite long and is a major factor controlling overall system performance.

The optimum width of a grounding strap made of copper sheet should be at least a tenth of its length. Therefore a 1.5 metre length copper sheet grounding strap should be at least 15 centimetres wide.

If, because of the size and construction of your boat, this guide cannot be complied with, then use the rule of thumb that the wider the grounding strap the better. In any case the strap should always be at least 50 mm wide (which is a common grounding strap dimension). The effectiveness of the grounding system on the HF installation should not be underestimated.

There should also be a minimum of wire connections in the grounding link. Significant effort should be made to ensure that the RF resistance in the link between the antenna tuner and the transceiver and the grounding strap is as low as possible.

As with all electrical connections, corrosion is the greatest cause of problems. It is therefore most important that the grounding component of the installation be kept in good condition and inspected regularly for signs of corrosion.

INTERFERENCE

Radio and Electrical Interference

Electrical and other radio or radar emissions, if originating from a source in close proximity to the HF antenna will cause interference to HF reception. To minimise this interference operators should ensure the antennas of other high power transmitting equipment are situated at the furthest distance possible from the HF antenna.

The power, control and antenna cabling of other electronic equipment in the vessel should be run separately from the cabling for the HF transceiver. This is particularly important for power cables which must not be run adjacent to the antenna cable.

HF radio transmissions generate large amounts of "inductive power" which is very easily picked up by other electronic equipment particularly auto pilots and fluxgate compasses. Great care should be taken when operating the HF transmitter whilst the vessel is being controlled by an auto pilot unit.

Interference from other electrical, radio and radar equipment may be checked by operating all other equipment individually and monitoring noise levels on the HF distress and safety frequencies. If unwanted noise (interference) is experienced, try to minimise it by separating the HF cabling from the cabling of the suspected interfering equipment. 12 Volts fluorescent lamps are another source of interference which should be considered; it is recommended that these lights be located away from the transceiver/tuner unit. Professional help can also be sought to fit suppressing equipment to the item of equipment causing the interference.

Non-electrical Aspects
Other aspects of a HF transceiver installation which need to be kept in mind are: position the transceiver so that surrounding noise sources do not interfere with reception; the transceiver should be sited in the deckhouse or cabin away from doorways and windows where it could be subjected to spray or rain; a remote speaker should be fitted so that calls may be monitored when operators may be working on deck and out of range of the transceiver the speaker should be protected from the weather.

CONCLUSION
Like most equipment, a HF installation will give optimum performance provided it has been installed correctly and is properly maintained. In this SEA, the HF installation has been broken down to its basic component parts. Each part has been discussed so that operators are aware of the effort required to achieve peak performance from the HF installation as a whole.

In the interest of marine safety, addressees are invited to reproduce this article and acknowledge the source.

Safety Operations Branch, Australian Maritime Safety Authority, PO Box 1108, BELCONNEN ACT 2616

SEPTEMBER 1991

APPENDIX 5

AUSREP
The contents of this Appendix reproduced by kind permission of the Australian Maritime Safety Authority, to whom grateful thanks is extended.

Australian Maritime
Safety Authority

AUSREP

The Australian Ship Reporting System

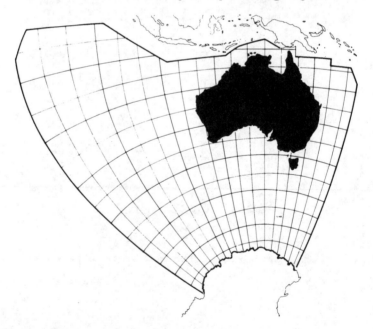

Maritime Rescue Coordination Centre

Co-ordinates of Australian SAR and AUSREP area are:

The coast of the Antarctic continent in longitude 75° E thence

6°00'S	75°00'E	9°37'S	141°02'E
2°00'S	78°00'E	9°08'S	143°53'E
2°00'S	92°00'E	9°24'S	144°13'E
12°00'S	107°00'E	12°00'S	144°00'E
12°00'S	123°20'E	12°00'S	155°00'E
9°20'S	126°50'E	14°00'S	155°00'E
7°00'S	135°00'E	14°00'S	161°00'E
9°50'S	139°40'E	17°40'S	163°00'E
9°50'S	141°00'E	to the coast of the Antarctic continent in longitude 163°00'E

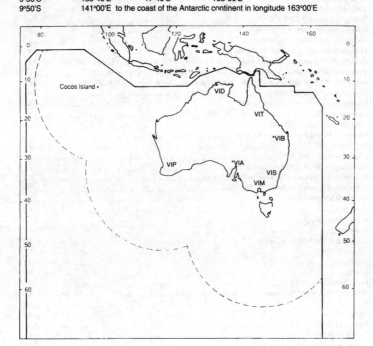

- - - Depicts approximate radius of action for Australian based long range search aircraft

* Note: VIA and VIB* available only until 31 January 1993

AUSTRALIAN SHIP REPORTING (AUSREP) AREA

145

Contents

Foreword

The Australian Ship Reporting System (AUSREP) has been established in accordance with the International Convention for the Safety of Life at Sea, 1974 (SOLAS) which requires signatories to the convention to provide marine search and rescue (SAR) facilities for prescribed areas, and also recommends the establishment of a ship reporting system. Australia, as a signatory to the 1974 SOLAS convention, has accepted SAR responsibility for the area shown in the attached general instructions and AUSREP procedures.

The AUSREP system was established in 1973 and is operated by the Maritime Rescue Coordination Centre, Canberra, which is part of the Australian Maritime Safety Authority. It is a 'positive' system that is, if a position report or final report is not received, the Maritime Rescue Coordination Centre will initiate checks to establish the safety of the vessel. These checks are aimed solely at establishing whether a vessel is safe, and include broadcasts to shipping and communications with owners, agents or charterers. If these checks are unsuccessful, then air search action will be initiated. As a 'positive' system it is important that Masters comply with the defined procedures as closely as circumstances permit.

Should you need any further information, please do not hesitate to contact the Maritime Rescue Coordination Centre, telephone (06) 247 5244.

May I take this opportunity to thank those Masters who have been regular participants of AUSREP and to welcome those who are new participants to the reporting system. I would also be grateful if those of you who have not done so, would complete and promptly forward the ship's particulars form located in Appendix B to this publication.

In conclusion, may I also make the point that the effectiveness of AUSREP and any SAR response depends on the accuracy and timeliness of your reports.

P.M. McGRATH
Chief Executive
Australian Maritime Safety Authority
August 1991

1.0 Functions of the Maritime Rescue Co-ordination Centre

The International Convention for the Safety of Life at Sea, 1974 (1974 SOLAS), and previous conventions, requires governments to '...ensure that any necessary arrangements are made for persons in distress at sea around its coasts'. The Australian Government, being a signatory to the 1974 SOLAS Convention, established the then Marine Operations Centre in April 1972.

The present Maritime Rescue Co-ordination Centre (MRCC) was previously known as the Sea Safety Centre and is now part of the Australian Maritime Safety Authority situated in Canberra in the Australian Capital Territory. This Authority was established on 1 January 1991.

The principal function of the MRCC is the co-ordination of marine search and rescue (SAR) activity, within the Australian area of responsibility which is shown on the inside cover. The MRCC is also part of the international COSPAS/SARSAT system and is equipped to receive and evaluate information transmitted by Emergency Position Indicating Radio Beacons (EPIRBS).

The MRCC is operated twenty-four hours a day by persons with considerable marine and in some cases aviation qualifications and experience. These officers have specialised knowledge of search and rescue procedures.

Communications are at the very heart of all search and rescue co-ordination and the MRCC is well equipped with a network enabling worldwide coverage. This network includes telephone, telex, facsimile, and satellite systems with radio communications provided by OTC Ltd, which maintains the maritime radio communications stations (MCS) network in Australia. These MCS provide the link between a ship and shore, and have telex lines to the MRCC.

An important part of the SAR framework is the Australian Ship Reporting System (AUSREP), which was introduced in December 1973 on a voluntary basis. In 1981 legislation was passed making AUSREP compulsory for all Australian-flag vessels, and for foreign vessels under the circumstances which are explained in the general instructions in the following pages. On voyages where it is not compulsory, ships are encouraged to participate whenever they are in the AUSREP area.

In addition to the MRCC's functions in the SAR role it is also the section of the Australian Maritime Safety Authority responsible for:

    (i) co-ordination and promulgation of Maritime Safety Information NAVAREA X warnings, which includes Australian Coastal Navigation Warnings (AUSCOAST).

    (ii) Operation of the Licensed Foreign Fishing Vessel Reporting System in the 200 nautical miles Australian Fishing Zone.

The facilities of the MRCC are available to all seafarers whether they are private boating enthusiasts, commercial fishermen or professional seamen of the merchant fleets of the world.

2.0 Australian Ship Reporting System General Instructions

2.1 Introduction

The Commonwealth of Australia Navigation Act 1912 (Division 14 Part IV) makes participation in the Australian Ship Reporting System (AUSREP) compulsory for certain ships. AUSREP is a ship reporting system established in accordance with the International Convention for the Safety of Life at Sea, 1974. Its objectives are:

(i)   to limit the time between the loss of a vessel and the initiation of search and rescue action, in cases where no distress signal is sent out;

(ii)   to limit the search area for a rescue action;

(iii) to provide up-to-date information on shipping resources available in the area, in the event of a search and rescue incident.

The coverage of AUSREP and the Australian search and rescue (AUSSAR) area, as advised to the International Maritime Organization (IMO), is illustrated, together with the co-ordinates of the area on the inside front cover.

2.2 Applicability of Navigation Act 1912 Division 14 Part IV

Division 14 Part IV of the Act apply to:

a. All Australian-registered ships engaged in interstate or overseas trade and commerce , while in the AUSREP area.

b. Ships not registered in Australia, but engaged in the coasting trade between Australia and an external territory, or between external territories, while in the AUSREP area.

c. Ships not registered in Australia, but demised under charter parties to charterers whose residences or principal places of business are in Australia, while in the AUSREP area.

d. Foreign ships, other than the above-mentioned vessels, from their arrival at their first Australian port until their departure from their final Australian port. However, they are encouraged to participate from their entry into and final departure from the AUSREP area.

e. Australian fishing vessels proceeding on overseas voyages, while in the AUSREP area, but not including those vessels operating from Queensland ports, which may call at ports in Papua New Guinea as an incidental part of their fishing operations. A definition of 'overseas voyage' is given in section 6 (1) of the Navigation Act 1912.

2.3 Offences

The Navigation Act 1912 provides penalties involving fines of up to $10 000 for infringements of reporting provisions.

2.4 Operating Authority

The Australian Maritime Safety Authority through the Maritime Rescue Coordination Centre,

Signal Address: MRCC Australia

Telephone: Canberra (06) 279 5916, (06) 279 5913, (06) 279 5914

Telex: 62349 (computer connected) - Answerback MRCC AUS

                Refer appendix D for INMARSAT A and INMARSAT C procedures

Facsimile: Canberra (06) 257 2036

2.5 Main Features of AUSREP

On departure from an Australian port or on entering the AUSREP area, a Sailing Plan (SP) is sent to the MRCC where a computerised plot is maintained of the vessels position. Position Reports (PR) are sent each day at the time selected by the Master so that a report is received at least every twenty four hours. *Should a vessel at any time be in a position more than two hours steaming from the position that would have been predicted from the last SP or PR, then a Deviation Report must be sent*. Failure to do so, in the event of a missed report, will result in the search being concentrated in the wrong area and the possibility that survivors from a stricken ship may not be found. On arrival at the ships destination or on departure from the AUSREP area a Final Report is to be sent. AUSREP is a positive reporting system which means that, should an expected report not be received, action including worldwide communication checks, alerting of ships in the vicinity and launching of search aircraft, will be initiated. Masters should note that in some parts of the AUSREP area the ability to conduct an air search may be restricted by aircraft range limitations (see chartlet inside front cover).

Dates and times contained in AUSREP reports are to be in Universal Coordinated Time which is indicated by the suffix 'UTC'.

2.6 Amendments and Alterations

The vertical unbroken line indicates information which has changed, been amended or is additional information since July 1989 when the previous edition of these instructions was published.

3.0 AUSREP Format Components

The following list shows all the AUSREP format components. The full list of components does not have to be included when sending AUSREP reports. Masters should include those components shown in the examples given for each report, others may be included at the Master's discretion or when relevant to the type of report being sent.

To assist with data processing, components of all reports should be preceded by the relevant alphabetical character and a stop (.)

The following format complies with the reporting format contained in IMO Resolution A648 (16) of 19 October 1989.

A. Vessel name and callsign

B. Date/Time of position (UTC)

C. Position (latitude and longitude)

D. Geographical position (not to be used in AUSREP messages)

E. Course

F. Speed (vessel's anticipated average speed until next report)

G. Name of last non-Australian port of call.

H. Date/Time (UTC) and point of entry into AUSREP system. The point of entry must be the Australian Port the vessel is departing from, or if entering the AUSREP area, the latitude/longitude of crossing the AUSREP boundary.

I. Next foreign (non-Australian) destination and estimated time of arrival at that port.

J. Whether pilot is carried on vessel (notification required on all Great Barrier Reef passages)

K. Date/Time (UTC) and point of exit from AUSREP system (point of exit is the latitude/longitude at which the vessel is leaving the AUSREP area, or the Australian port the vessel is to arrive)

L. Route (vessel's intended track—state Rhumb Line or Coastal Great Circle or Composite with Limiting Latitude)

M. Coast radio maritime communications stations monitored (include INMARSAT A&C numbers if fitted)

N. Daily reporting time

0. Draught

*P. Cargo

*Q. Defects or other limitations

*R. Pollution (or reports of any pollution seen)

S. Weather conditions in area

T. Ship' s agents

U. Ship type and size

V. Medical personnel carried (Sailing Plan only)

W. Number of persons on board

X. Remarks

*Harmful substances, Marine Pollution, Dangerous Goods and probability of discharge reports refer Appendix A.

4.0 Types of Reports

4.1 Sailing Plan Report (AUSREP SP)

A sailing plan report is sent to the MRCC either within twenty-four hours prior to or up to two hours after entry into the AUSREP area or departure from a port within the AUSREP area.

The SP contains information necessary to initiate a plot and gives an outline of the intended passage.

Should a vessel not sail within two hours of the time stated in the SP the plan should be cancelled and a new SP sent.

If the Master of a foreign vessel departing on an overseas voyage from an Australian port does not intend sending position reports, this must be indicated in the SP by the inclusion of the words NOREP in place of the daily reporting time. Under this option the MRCC will not undertake search and rescue action unless specific information is received which indicates an air search is warranted.

4.2 Commencement of Active Sar Watch

Masters should be aware that if a vessel lodges an SP prior to entering the AUSREP area, active SAR watch on their vessel WILL NOT COMMENCE until a report is received by the Centre indicating the vessel has entered the AUSREP area.

4.3 Mandatory AUSREP SP Components

AUSREP SP

A, F, H, K, L, M, N, V.

Notes:

(i) refer page 3 for components;

(ii) include component 'G' when entering AUSREP from overseas, or component 'I' when leaving AUSREP to an overseas destination;

(iii) if ship intends to transit Great Barrier Reef/Torres Strait notification is required regarding the carriage of a Pilot (see paragraph J previous page);

(iv) other sections of the basic format may be included at Masters discretion: for example 'X' if report required to be passed AMVER, or details of name/callsign change since last report.

4.4 Example of SP on Entering AUSREP Area

Format	Example
AUSREP SP	AUSREP SP
A. Vessel name and callsign	A. HESPERUS/BCBC
F. Speed	F. 12
G. Name of last port of call (when entering from overseas)	G. Port Said
H. Date/Time and point of entry into AUSREP system	H. 020400UTC 0448S 07555E
K. Date/Time and point of exit from AUSREP system (Australian port of arrival)	K. Adelaide 180600UTC
L. Route (vessels intended track)	L. 129 to Cape Leeuwin then direct coastal route
M. Coast radio/maritime communications stations monitored (Marisat number if fitted)	M. GKA, 9VG, VIP, VIS (INMARSAT 1290617)
N. Daily reporting time	N. 0600 (UTC)
V. Medical personnel	V. No medic
X. Remarks	X. Pass to AMVER, name/callsign changed from SWEETAPPLE/ABAB since last report.

Example of Message Transmitted: AUSREP SP **A.** HESPERUS/BCBC **F.** 12 **G.** Pt Said **H.** 020400UTC 0448S 07555E **I.** Adelaide 180600UTC **L.** 129 to Cape Leeuwin **M.** GKA VIP VIS INMARSAT 1290617 **N.** 0600UTC **V.** no medic **X.** pass to AMVER, name/callsign changed from SWEETAPPLE/ABAB since last report.

- 4 -

151

4.5 Example of SP submitted on departure from a Port within AUSREP Area to a Port outside AUSREP Area
(May be submitted up to twenty-four hours prior to sailing and up to two hours after departure)

Format	Example
AUSREP SP	AUSREP SP
A. Vessel's name and callsign	A. HESPERUS/BCBC
F. Speed	F. 1 2
H. Date/Time and departure port for this voyage	H. 020500UTC Fremantle
I. Next overseas destination and ETA	I. Jakarta 090600UTC
K. Date/Time and point of exit from AUSREP area	K. 080600UTC 1200S 10700E
L. Route (vessels intended track)	L. To 2900S 11327E then 1200S 10700E
M. Coast radio/maritime communications stations monitored	M. GKA, VIP, V10, PRI, 9VG. (INMARSAT 1290617)
N. Daily report time	N. 0400UTC
V. Medical personnel	V. No medic
X. Remarks	X. Report to AMVER

Example of Message Transmitted: AUSREP SP **A**. HESPERUS/BCBC **F**. 12 **H**. 020500UTC Fremantle **I**. 090600UTC Jakarta **K**. 080600UTC 1200S 10700E **L**. to 2900S 11327E to 1200S 10700E **M**. GKA VIP 9VG (INMARSAT 1290617) **N**. 0400UTC **V**. no medic **X**. pass to AMVER

Note: In the case of a foreign vessel whose next port of call is outside Australia, if the Master does not intend to send daily position reports, this must be indicated in the AUSREP SP by inclusion of the words NOREP in paragraph N.

4.6 Example of SP submitted on departure from and to a Port both within AUSREP Area
(May be submitted up to twenty-four hours prior to sailing and up to two hours after departure)

Format	Example
AUSREP SP	AUSREP SP
A. Vessel's name and callsign	A. HESPERUS/BCBC
F. Speed	F. 12
H. Date/Time and point of entry into AUSREP system (Australian port of departure)	H. 212200UTC Adelaide
K. Date/time and point of exit from AUSREP system (Australian port of arrival)	K. Melbourne 231400UTC
L. Route (Vessel's intended track)	L. Coastal direct
M. Coast radio/maritime communications station monitored	M. Via VIM
N. Daily reporting time	N. 0100UTC
V. No medic	V. Medical personnel
X. Nil	X. Remarks

Example of Message Transmitted: AUSREP SP **A**. HESPERUS/BCBC **F**. 12 **H**. 212200UTC Adelaide **I**. Melbourne 231400UTC **L**. Coastal direct **M**. VIM N. 0100UTC **V**. no medic

- 5 -

152

4.7 Example of SP submitted when transitting AUSREP Area from and to a Port not in AUSREP Area

Format	Example
AUSREP SP	AUSREP SP
A. Vessel's name and call sign	A. HESPERUS/BCBC
F. Speed	F. 15
G. Name of last port of call	G. Durban
H. Date/time and point of entry into AUSREP system	H. 120800UTC 3550S 07500E
I. Next overseas destination and estimated time arrival	I. Surabaya 190200UTC
K. Date/time and point of exit from AUSREP system	K. 172330UTC 0806S 10107E
L. Route	L. Great Circle
M. Coast radio/maritime stations monitored	M. VIP, 9VG, PKI
N. Daily report time	N. 0600UTC
V. Medical personnel carried	V. No medic
X. Remarks	X. Pass to AMVER

Example of Message Transmitted: AUSREP SP **A**. HESPERUS/BCBC **F**. 15 **G**. Durban **H**. 120800UTC 3550S 07500E **I**. Surabaya 190200UTC **K**. 172330UTC 0806S 10107E **L**. RL **M**. VIP 9VG PKI **N**. 0600UTC no medic **X**. pass to AMVER

4.8 Position Report (AUSREP PR)

Each day at the nominated daily reporting time given in the SP, a PR should be transmitted to the MRCC. The first PR is required within 24 hours of the SP, and daily thereafter at the nominated daily reporting time, including the day of arrival at, or departure from the AUSREP area. The information contained in the PR will be used by the Centre to update the plot. The PR must reflect the position, course and speed of the ship at the time of the report. If it is necessary to alter the nominated daily reporting time, the alteration should be shown in the PR sent before the change.

If any PR is sent at a time other than previously nominated by the ship, the time of this PR must not exceed 24 hours from the previous report.

The ETA at the Australian destination, or AUSREP area boundary, must be confirmed in the last PR of a passage. It may also be amended in any report whenever the Master is aware of a revised ETA.

4.9 Mandatory AUSREP components for a PR

Mandatory AUSREP format components for a PR are:
AUSREP PR
A, B, C, E, F.

4.10 Example of a PR

Format	Example
AUSREP PR	AUSREP PR
A. Vessel's name and callsign	A. HESPERUS/BCBC
B. Date/time of position (UTC)	B. 030400UTC
C. Position	C. 0748S 07940E
E. Course	E. 129
F. Speed	F. 8.5
X. Remarks (for example, change of reporting time or revised ETA)	X. ETA now 060200UTC

Example of Message Transmitted: AUSREP PR **A**. HESPERUS/BCBC **B**. 030400UTC **C**. 0748S 07940E **E**. 129 **F**. 8.5 **X**. ETA now 060200UTC

- 6 -

4.11 Deviation Report (AUSREP DR)

Should a vessel, at any time, be in a position more than two hours steaming from the position that would be predicted from the last SP or PR, then a deviation report **MUST BE SENT**. Masters are reminded that this is an important component of the system in that any alteration of course or speed sufficient to affect the estimated position as above may result in the search effort being concentrated in the wrong area should the vessel subsequently be in distress.

4.12 Mandatory AUSREP components for a DR

Mandatory AUSREP format components for a DR are:
AUSREP DR
A, B, C + (extra components-see example below).

4.13 Example of a DR

Format	Example
AUSREP DR	AUSREP DR
A. Vessel name and call sign	A. HESPERUS/BCBC
B. Date/time of position (UTC)	B. 050200UTC
C. Position	C. 3900S 14500E
(Then include whichever format components are affected by the deviation)	
F. Speed	F. 9
I. Destination and ETA	I. Adelaide 080200UTC
X. Remarks (include reason for deviation, for example, reduction in speed due weather, change of route or port of destination etc.)	X. Reduced speed due to main engine problems

Example of Message Transmitted: AUSREP DR **A.** HESPERUS/BCBC **B.** 050200UTC **C.** 3900S 14500E **F.** 9 **I.** Adelaide 080200UTC **X.** reduced speed due to main engine problems

4.14 Final Report (AUSREP FR)

When the vessel approaches the Australian destination and arrives at a position where VHF contact is made with the local harbour authority or pilot station, which under normal conditions is within two hours steaming of the pilotage, a FR is sent to the MRCC. Under no circumstances should a FR be sent more than two hours before arrival. Alternatively, if the arrival is outside radio watchkeeping hours, the FR may be phoned immediately after berthing, but no later than two hours after arrival. If it is known that the vessel is to anchor or berth where telephone facilities are not available, then the FR should be passed through the appropriate coast radio/maritime communications station.

For a vessel departing from the AUSREP area the FR is to be sent after crossing the area boundary. Again, Masters are requested to ensure that an FR is sent to the MRCC, to prevent unnecessary search action.

4.15 Mandatory AUSREP components for an FR

Mandatory AUSREP format components for an FR are:
AUSREP FR
A, K, X.

4.16 Example of an FR on arrival at a port in AUSREP area

Format	Example
AUSREP FR	AUSREP FR
A. Vessels name and callsign	A. HESPERUS/BCBC
K. Date/Time and port of arrival	K. 080200UTC Adelaide
X. Remarks (must include the words final report)	X. final report

Example of Message Transmitted: AUSREP FR **A**. HESPERUS/BCBC **K**. 080200UTC Adelaide **X**. Final Report

4.17 Example of FR, vessel departing AUSREP area

Format	Example
AUSREP FR	AUSREP FR
A. Vessels name and callsign	A. HESPERUS/BCBC
K. Date/Time and point of exit from AUSREP area	K. 070030UTC 1100S 10631 E
X. Remarks (must include the words final report)	X. final report

Example of Message Transmitted: AUSREP FR **A**. HESPERUS/BCBC **B** 070030UTC 1100S 10631E **X**. Final Report

4.18 Method of Passing Reports

In an Australian port
A reverse charge telephone call or telex may be used when making an AUSREP or SAR message.Should INMARSAT A or C be used refer to Appendix D, (Page 18)

It is advisable to communicate all reports direct from the ship to the MRCC to avoid delays associated with intermediate agencies: telephone (06) 279 5916 (reverse charge); Telex 62349; Facsimile (06) 257 2036.

At sea
All reports are addressed MARITIME RESCUE COORDINATION CENTRE CANBERRA and may be sent free of charge through any Australian maritime communication station controlled by OTC Ltd. Schedules and frequencies are given in Admiralty List of Radio Signals, Vol. 1. Reports passed through INMARSAT will be free of charge providing procedures outlined in Appendix D, (Page 18) are used. All reports sent by voice should include the mandatory components including the identifying letter.

5.0 Overdue Reports

TO AVOID UNNECESSARY SEARCH ACTION IT IS MOST IMPORTANT that ships report at their reporting time each day and send their FINAL REPORT when leaving the AUSREP area. If a ship is unable to pass a position report for any reason (for example, unserviceable radio equipment) attempts must be made to pass a signal to this effect through another vessel, harbour or other shore authority, either by VHF, signalling lamp, or use of emergency transmitter. Masters are requested to ensure that these procedures are followed. Please remember the AUSREP system is in operation FOR YOUR SAFETY.

Action taken by the MRCC

As AUSREP is a positive reporting system; if an expected report is not received by the MRCC within two hours, necessary steps are taken to ascertain the safety of the vessel. The following is an outline of the action that will be taken should a report become overdue. Circumstances may dictate more rapid action.

1. During the first two hours internal checks will be carried out.
2. Vessel will be listed on traffic lists requesting master to furnish the overdue report.
3. At six hours overdue broadcast of ships callsign with (JJJ/REPORT IMMEDIATE indicator*) will precede traffic lists indicating concern due to non-receipt of PR/FR. An all station (CQ) inquiry may be initiated.

- 8 -

4. Extensive communication checks with Australian and overseas radio stations, owners, agents and other vessels are carried out to trace the last sighting or a contact with the vessel with the aim of confirming its safety.

5. At twenty-one hours overdue the JJJ/REPORT IMMEDIATE broadcast may be upgraded to the Urgency Signal XXX/PAN indicator. Search planning will be in progress and details included in NAVAREA X and facsimile weather broadcasts via AXM and AXI. By the time the report is twenty-four hours overdue, positive action will have been initiated to locate the vessel. This action will include the launching of search aircraft** where possible. However, as indicated in paragraph 2.5, due to aircraft range limitations the resources available for an air search decrease with distance from an Australian base.

* This signal consists of the vessel's call sign followed by JJJ/ (RTG) or REPORT IMMEDIATE (RTF). Any sighting of, or communication with, this vessel by any other vessel should be reported immediately to the CRS/MCS stating time of contact, position and estimated course and speed of the vessel overdue for report.

** Search aircraft will not necessarily be launched when it is known that the vessel is equipped with a 406 MHz float free EPIRB

6.0 Notes on Procedures

6.1 Date/Time Group

Dates and times contained in AUSREP reports are to be in Universal Co-ordinated Time (UTC) which is indicated by the suffix 'UTC'.

Example: 5 November 1982, 0600LMT (UTC 9 hrs) transmitted as 042100UTC.

6.2 Latitude and Longitude

Latitude
Four figure group indicated by suffix 'S' (south)
Example: Latitude 06°15'46" south. Transmitted at 0616S.
Longitude
Five figure group indicated by suffix 'E' (east)
Example: Longitude 82°06'24" east. Transmitted as 08206E.

6.3 Intended Route

Indicate Great Circle or Rhumb Line with way points being followed, expressed in latitude or longitude. Courses are not required if way points are mentioned.

6.4 Speed

Anticipated average speed vessel will make till next report time.

6.5 Course

True course anticipated until next reporting time. When more than one course will be steered enter 'various'. If course shown as 'various', the Centre will interpret this as being the normal courses a vessel will follow on that particular passage. However, if the above assumption does not apply, clarification should be made.

6.6 Radio Stations Monitored

Details should be provided of those radio stations which the vessel normally works for commercial radio communications. in addition to the Australian maritime communications stations monitored. If the ship is fitted with INMARSAT, the station identity number(s) should also be included.

6.7 Daily Reporting Time

When selecting their daily reporting time, Masters of single-radio-operator-ships, will appreciate that reports nominated during the last single-operator-period of the day, if not received, cannot be followed up until the next radio watch period. This may cause considerable delay in activating any SAR action that may be required. It is therefore desirable that reports are transmitted during the first single-operator-period of the day.
If any position report is different from the PR time, the next report should be no more than twenty-four hours from that PR time.

- 9 -

6.8 Two Hour Rule

Should a vessel, at any time, be in a position more than two hours steaming from the position that would be predicted from the last SP or PR, a deviation report **MUST BE SENT.**

6.9 Severe Weather

Vessels which anticipate passing through areas of severe weather conditions or are experiencing severe weather, are urged to report their position at more frequent intervals to the MRCC. The time of their next anticipated position report should be included under Remarks.

6.10 Reports to AMVER

Whilst participating in AUSREP, Masters may also wish their reports to be forwarded to New York for inclusion in the AMVER plot. This should be indicated in report remarks, and will particularly relate when sending the final report when the vessel leaves the AUSREP area.

Masters are requested to note that AMVER reports will only be forwarded to the US Coast Guard if the vessel is in the AUSREP area of responsibility, reporting to the AUSREP system. If the ship also requires its AUSREP report to be passed to AMVER a request is to be included in the report.

6.11 Reports to JASREP

Reports from ships to JASREP are not normally forwarded by the MRCC. Ships are requested to pass these reports direct.

6.12 Computer Processing at the MRCC

To enable efficient processing by the MRCC's computer, ships are encouraged to send reports in IMO format as detailed throughout this publication. Components should be clearly identified by the relevant alphabetical letter followed by a stop (.) and be listed across rather than down the page, this is particularly relevant to those vessels reporting by automatic or semi-automatic means. Please note examples of messages transmitted.

APPENDIX A: Detailed Reporting Requirements

A.1 Dangerous goods reports (DG)

Primary reports should contain items A, B, C, M, Q, R, S, T, U of the standard reporting format; details for R should be as follows:

R 1. Correct technical name or names of goods.
   2. UN number or numbers.
   3. IMO hazard class or classes.
   4. Names of manufacturers of goods when known, or consignee or consignor.
   5. Types of packages including identification marks. Specify whether portable tank or tank vehicle, or whether vehicle or freight container or other cargo transport unit containing packages. Include official registration marks and numbers assigned to the unit.
   6. An estimate of the quantity and likely condition of the goods.
   7. Whether lost goods floated or sank.
   8. Whether loss is continuing.
   9. Cause of loss.

If the condition of the ship is such that there is danger of further loss of packaged dangerous goods into the sea, items P and Q of the standard reporting format should be reported; details for P should be as follows:

P 1. Correct technical name or names of goods.
   2. UN number or numbers.
   3. IMO hazard class or classes.
   4. Names of manufacturers of goods when known, or consignee or consignor.
   5. Types of packages including identification marks. Specify whether portable tank or tank vehicle, or whether vehicle or freight container or other cargo transport unit containing packages. Include official registration marks and numbers assigned to the unit.
   6. An estimate of the quantity and likely condition of the goods.

Particulars not immediately available should be inserted in a supplementary message or messages.

A.2 Harmful substances reports (HS)

In the case of actual discharge primary harmful substances reports should contain items A, B, C, E, F, L, M, N, Q, R, S, T, U, X of the standard reporting format. In the case of probably discharge, item B should also be included. Details for P, Q, R, T and X should be as follows:

P 1. Type of oil or the correct technical name of the noxious liquid substances on board.
   2. UN number or numbers.
   3. Pollution category (A, B, C or D), for noxious liquid substances.
   4. Names of manufacturers of substances, if appropriate, where they are known, or consignee or consignor.
   5. Quantity.

Q 1. Condition of the ship as relevant.
   2. Ability to transfer cargo-ballast-fuel.

R 1. Type of oil or the correct technical name of the noxious liquid discharged into the sea.
   2. UN number or numbers.
   3. Pollution category (A, B, C or D), for noxious liquid substances.
   4. Names of manufacturers of substances, if appropriate, where they are known, or consignee or consignor.
   5. An estimate of the quantity of the substances.
   6. Whether lost substances floated or sank.
   7. Whether loss is continuing.
   8. Cause of loss.
   9. Estimate of movement of the discharge or lost substances, giving current conditions if known.
  10. Estimate of the surface area of the spill if possible.

T 1. Name, address, telex and telephone number of the ship's owner and representative (charterer, manager or operator of the ship or their agent).

X   1. Actions being taken with regard to the discharge and the movement of the ship.

    2. Assistance or salvage efforts which have been requested or which have been provided by others.

    3. The master of an assisting or salvaging ship should report the particulars of the action undertaken or planned.

Particulars not immediately available should be inserted in a supplementary message or messages.

The Master of any ship engaged in, or requested to engage in an operation to render assistance or undertake salvage should report, as far as practicable, items A, B, C, E, F, L, M, N, P, Q, R, S, T, U, X of the standard reporting format.

A.3 Marine pollutants reports (MP)

In the case of actual discharges, primary MP reports should contain items A, B, C, M, Q, R, S, T, U, X of the standard reporting format. In the case of probable discharge, item P should also be included. Details of P, Q, R, T and X should be as follows:

P   1. Correct technical name or names of goods.

    2. UN number or numbers.

    3. IMO hazard class or classes.

    4. Names of manufacturers of goods when known, or consignee or consignor.

    5. Types of packages including identification marks or whether in portable tank or tank vehicle or whether vehicle of freight container or other cargo transport unit containing packages. Include official registration marks and numbers assigned to the unit.

    6. An estimate of the quantity and likely condition of the goods.

Q   1. Condition of the ship as relevant.

    2. Ability to transfer cargo-ballast-fuel.

R   1. Correct technical name or names of goods.

    2. UN number or numbers.

    3. IMO hazard class or classes.

    4. Names of manufacturers of goods when known, or consignee or consignor.

    5. Types of packages including identification marks, specify whether in portable tank or tank vehicle or whether vehicle of freight container or other cargo transport unit containing packages. Include official registration marks and numbers assigned to the unit.

    6. An estimate of the quantity and likely condition of the goods.

    7. Whether lost goods floated or sank.

    8. Whether loss is continuing.

    9. Cause of loss.

T   1. Name, address, telex and telephone number of the ship's owner and representative (charterer, manager or operator of the ship or their agent).

X   1. Action being taken with regard to the discharge and movement of the ship.

    2. Assistance or salvage efforts which have been requested or which have been provided by others.

    3. The Master of an assisting or salvaging ship should report the particulars of the action undertaken or planned.

Particulars not immediately available should be inserted in a supplementary message or messages.

The master of any ship engaged in or requested to engage in an operation to render assistance or undertake salvage should report, as far as practicable, items A, B, C, M, P, Q, R, S, T, U, X of the standard reporting format.

- 12 -

A.4 Probability of discharge

The probability of discharge resulting from damage to the ship or its equipment is a reason for making a report. In judging whether there is such a probability and whether the report should be made, the following factors, among others, should be taken into account:

1. the nature of the damage, failure or breakdown of the ship, machinery or equipment; and

2. sea and wind state and also traffic density in the area at the time and place of the incident.

It is recognised that it would be impracticable to lay down precise definitions of all types of incidents involving probable discharge which would warrant an obligation to report. Nevertheless, as a general guideline the Master of the ship should make reports in cases of:

1. damage, failure or breakdown which affects the safety of the ships; examples of such incidents are collision, grounding, fire, explosion, structural failure, flooding, cargo shifting; and

2. failure or breakdown of machinery or equipment which results in impairment of the safety of navigation; examples of such incidents are failure or breakdown of steering gear, propulsion plant, electrical generating system, essential shipborne navigational aids.

Ship's Particulars — Australian Ship Reporting System (AUSREP)

Information contained in this form will only be used for the purpose of safety of life at sea.

Please attach two colour photographs of your vessel; one beam on, the other from the air (if possible).
If you are unable to supply photographs, please sketch the vessel - illustrating main features and colours.

Australian Maritime Safety Authority

Basic Vessel Details

Vessel's Name	Black Purple
Call Sign	Cyan
Port of Registry	Quebec
Previous Name	White Red
Previous Call Sign	Grey
Vessel Type	Bulk Carrier
Vessel Size (Length and Deadweight Tonnage)	185 metres 35000 DWT
Hull Colour	Black
Deck Colour	Orange
Superstructure Colour and Position	White/Aft
Normal Complement of Vessel	26

Doctor or Medical Personnel Carried Yes ☐ No ☐

Owner Operator Details

Registered Owner's Name	Orange White Inc.
Telephone (Business) International Code	616) 99-9999
Telex International Code	616) 623445
Facsimile International Code	616) 1517892
Address	999 Black Street Sydney NSW 2699 Australia

Complete this section if Registered Owner is NOT the actual vessel operator

Management Company's Name	Green Ship Management
Telephone (Business)	Sydney (02) 88-8888
Telex	8888
Facsimile	(02) 22-2222
Address	999 White Street Sydney NSW 2688 Australia

Please provide details of TWO contacts ashore (e.g. Marine Superintendents) who could supply current information on the vessel

Name	Mr Black
Telephone (Business)	Sydney (02) 22-2222
Telephone (Home)	Sydney (02) 33-3333
Address	888 Grey Street Sydney NSW 2677 Australia
Name	Mr White
Telephone (Business)	Sydney (03) 44-4444
Telephone (Home)	Sydney (02) 55-5555
Address	777 Blue Street Sydney NSW 2655 Australia

Australian Agent

Agent's Name	Blue Agency
Telephone (Business)	Melbourne (03) 44-4444
Telex	4444
Facsimile	(03) 77-7777
Address	999 Yellow Street Melbourne Vic 2599 Australia

Ship Classification Society

Classification Society's Name	Lloyd's Register
Telephone (Business)	London 99-9999
Telex	9999
Facsimile	88-8888
Address	333 Indigo St London U.K.
Classification Society Representative's Name	Lloyd's Agent
Telephone (Business)	Sydney (02) 66-6666
Telex	6666
Facsimile	33-3333
Address	777 Violet Street Sydney NSW 2655

Emergency Position Indicating Radio Beacons (EPIRBs)

121.5 EPIRBS

Quantity	Frequencies	Manufacturer	Where Stored	Operating Duration
2	121.5/243 MHz	Black Electronics Sydney	Liferafts	48 Hrs
2	121.5/406 MHz	White Electronics Perth	Bridge Wings	48 Hrs

406 EPIRB/L Band EPIRB/Ship Station Identity

Quantity	SSI & Band	Hex I.D. (406 MHz)	Serial No.	Country of Registration
1		BEED 0002000001	2001	Canada
1	351724513		4357/A	Canada

Radar Characteristics

Type	Frequency MHz	Pulse Repetition Frequency (MHz)	Pulse Width (milliseconds)	Scan Rate (seconds)
Decca 1226	9400	850/1700/3400/750	1.0/0.25	2
Ratheon 1450	3100	750	0.4	4

Search and Rescue Radar Transponder (SART)

Number	Frequency	Manufacturer	Where Carried
2	9.3 Giga Hz	Nippon Electronics Tokyo	Lifeboats

Position Fixing Aids Fitted (e.g. SATNAV) - Please list below

SatNav, DF, Loran, Radar, Omega, Decca

INMARSAT Numbers

INMARSAT A 1730315	Alternate 1730317	INMARSAT B 173031951	Alternate 173032052
INMARSAT C 1730321	Alternate 1730322	INMARSAT M 173032323	Alternate 173032414

D.S.C. Identification / SELCALL Number

D.S.C. Identification	SELCALL Number
503 000031	12151

Helicopter Position

Is there a helicopter transfer position? Yes ✓ No ☐

Is a helicopter able to land on this position? Yes ✓ No ☐

Where is position situated? No. 4 Hatch

What is the maximum size/weight deck can support? Size 100 SQ metres Weight 30

Please attach a list of any other details you consider relevant or which may assist search and rescue co-ordinators.

Date 26th January 1991

- 14 -

161

Appendix C: Check list for AUSREP reports

AUSREP REPORTS CHECK LIST

—	SP	PR	FR	DR	
A	*	*	*	*	name/call sign
B		*		*	date/time of position
C		*		*	lat / long of position
E		*		†	course
F	*	*		†	speed
G	‡				last port of call, only when entering from overseas
H	*				date/time and point of entry (lat/long) into AUSREP area or Australian port of departure
I	‡			†	next overseas destination and ETA
J	§				whether pilot is carried on vessel
K	*		*	†	date/time and point of exit, either the next Australian port OR lat/long when leaving the AUSREP area.
L	*			†	route
M	*			†	MCS monitored/INMARSAT NUMBER
N	*			†	nominated daily reporting time
V	*				medical personnel carried
X			*	†	remarks

*   Mandatory

‡   this information is only required when entering or departing the AUSREP area

†   Include these only if affected by the deviation

§   Notification of pilot required for all Barrier Reef passages

More detailed information about these message components will be found on page 3

Appendix D

INMARSAT A, INMARSAT C Procedures

The following applies for the forwarding of AUSREP reports by INMARSAT from Pacific and Indian Ocean regions to Perth Coast Earth Station (CES). These procedures apply strictly to AUSREP messages only. Ships will not be charged for these messages if correct procedures are followed.

INMARSAT A fitted Ship Earth Stations

Select: * a. 02 (for Perth) then

       b. service code 43+

* Ensure ship's antennae is directed at appropriate Indian or Pacific Ocean satellite.

INMARSAT C fitted Ship Earth Stations

(The following may not apply to early versions of INMARSAT Standard C software)

Select: a (Destination) 7162025

       b (Coaststation Perth) 202

notes:

(i) There could be delays in Perth Coast Earth Station's store and forward facility, particularly during busy periods. If ships become concerned as to delivery to the MRCC of AUSREP messages they should first check the STATUS of the message as displayed on their Standard C Mobile Ship Earth Station (SES). STATUS provides information when a (CES) has first received the message then confirms the message has been delivered to the addressee.

If **no** confirmation of delivery has been received some 30 minutes after receiving STATUS information indicating the transmission has been acknowledged by the CES but not delivered to the MRCC, ships should then telex LES Perth (71197075), to inquire as to the status of their AUSREP messages before attempting to send message again.

(ii) Whilst reporting to AUSREP ships are to ensure their INMARSAT equipment remains active in the "LOGIN" mode

(iii) At the time of publishing INMARSAT were developing a short data report for INMARSAT C type systems. This enables transmission of certain messages at much reduced cost. Information will be promulgated when this system has been satisfactorily developed and implemented.

BIBLIOGRAPHY

1. Global Maritime Distress and Safety System
    IMO London 1987 (ISBN 92-801-1216-3)

2. Never Beyond Reach
    INMARSAT London 1989

3. INMARSAT Users Handbook
    INMARSAT London

4. NAVTEX Manual IMO London 1968
    (ISBN 92-801-1238-4)

5. Final Acts of the World Administrative Radio Conference for the
    Mobile Services (MOB-87)
    ITU Geneva 1988
    (ISBN 92-61-03101-3)

6. Appendix 11 and Article 26 to the World Administrative Radio
    Conference for the Mobile Services (MOB-87)
    ITU Geneva 1987

7. Final Acts of the World Administrative Radio Conference for the
    Mobile Services (MOB-83)
    ITU Geneva 1983
    (ISBN 92-61-01731-2)

8. Manual for Use by the Maritime Mobile and Maritime Mobile
    Satellite Services.
    ITU Geneva 1991.

9. Amendments to the 1974 SOLAS Convention Concerning
    Radiocommunications for the Global Maritime Distress and
    Safety System.
    IMO London 1989.
    (ISBN 92-801-1249-X)

10. Amendments to the Protocol of 1978 relating to the International
    Convention for the Safety Of Life At Sea 1974 Concerning
    Radiocommunications for the Global Maritime Distress and
    Safety System
    IMO London, 1989
    (ISBN 92-801-1250-3)

11. International Convention for the Safety Of Life At Sea

IMO London 1986 consolidated edition.
(ISBN 92-801-1200-7)

12. International Conference on Training and Certification of
    Seafarers.
    IMO London 1978
    (ISBN 92-801-1085-3)

13. International Code of Signals
    IMO London 1985 1987 (with Supplement 1989)
    (ISBN 92-801-1184-1)

14. Merchant Ship Search and Rescue Manual.
    IMO London 1986
    ISBN 92-801-1205-2)

15. The International Hydrographic Review
    International Hydrographic Bureau Monaco
    (Journal January 1989)

16. WARC MOB (87) Australian DOTC Delegation Report of the
    Australian
    Delegation to the ITU World Administrative Radio Conference
    for the Mobile Services (Geneva, 1987).
    Department of Transport and Communications, Canberra, 1987.

17. Information Paper on the Global Maritime Distress and Safety
    System (GMDSS) for the Australian Shipping Industry
    Department of Transport and Communications Canberra, June
    1990.

18. Implementation of the Global Maritime Distress and Safety
    System by Australia Implications for Fishing Vessels and Pleasure
    Craft Department of Transport and Communications, Canberra
    May 1989

19. GMDSS General Operator's Certificate Handbook for Australian
    Vessels (Revised April 1991)
    Department of Transport and Communications Canberra

20. Ocean Voice 'INMARSAT's quarterly journal
    INMARSAT, 40 Melton Street, London NW1 2EQ
    (ISSN 0261-6777)

21. Safety at Sea
International Trade Publications, Redhill, U.K. Monthly journal
(ISBN 0142-0666)

22. Global Maritime Distress and Safety System – General Operator's
Certificate – Course and Assessment Information Australian
Maritime College. October 1991

23. Consolidated Text of the 1974 SOLAS Convention the 1978
SOLAS Protocol the 1981 and 1983 SOLAS Amendments IMO
London 1986

24. Recommendation 493-3 (MOD I) from Conclusions of the
Interim Meeting of Study Group 8 – Maritime Mobile Service
CCIR Study Groups 1986-1990

25. Admiralty List of Radio Signals (Volumes) 1985 The
Hydrographer of the Navy, Taunton, Somerset. U.K.

26. MANUAL for use by the Maritime Mobile and Maritime Mobile-
Satellite Services Geneva 1992

27. The GMDSS Handbook IMO London 1992

28. INMARSAT-C Maritime User's Manual INMARSAT London
1992

29. IPS RADIO AND SPACE SERVICES USER TRAINING
MANUAL Australian Department of Administrative Services
1990

30. FINAL ACTS OF THE REGIONAL ADMINISTRATIVE
RADIO CONFERENCE FOR THE PLANNING OF THE MF
MARITIME MOBILE AND AERONAUTICAL
RADIONAVIGATION SERVICES (REGION 1) (GENEVA
1985) ITU Geneva

BROCHURES:-

INMARSAT-C Communications System – Abstract
Kevin Phillips

INMARSAT-C
INMARSAT 1990

Land Mobile and Special Services for INMARSAT
INMARSAT

Satellite Communications on the Move
INMARSAT October 1989

Alive via Satellite – Brochure and video.
Department of Transport and Communications Canberra January 1990.

Changes Ahead – Maritime Safety Communications – Brochure.
Department of Transport and Communications Canberra.
February 1990.

Preliminary Details for Small Ships on the new Global Maritime
Distress and Safety System
OTC Maritime September 1989

Details of High Frequency Changes on 1st July 1991
OTC Maritime March 1991

OTC Satcoms – A Revolution in Mobile Communications

Search and Rescue in Australia – the satellite era
Department of Transport and Communications, Canberra, January 1991

Satellite Compatible EPIRBs

Australian Maritime Safety Authority, Canberra , January 1991

INMARSAT 'Facts' Brochures:-

February	1987 – Standard-C
March	1988 – Global Maritime Distress and Safety System
April	1988 – Standard-C
June	1989 – Enhanced Voice Group Call
July	1989 – Land Mobile Satellite Communications
December	1989 – Global Maritime Distress and Safety System
December	1989 – INMARSAT-C.

Provision and Procedure for Distress and Safety Calls
Annex 1 to Report 747.

INDEX

172